Quantitative Operational Risk Models

CHAPMAN & HALL/CRC FINANCE SERIES

Aims and Scopes

As the vast field of finance continues to rapidly expand, it becomes increasingly important to present the latest research and applications to academics, practitioners, and students in the field.

An active and timely forum for both traditional and modern developments in the financial sector, this finance series aims to promote the whole spectrum of traditional and classic disciplines in banking and money, general finance and investments (economics, econometrics, corporate finance and valuation, treasury management, and asset and liability management), mergers and acquisitions, insurance, tax and accounting, and compliance and regulatory issues. The series also captures new and modern developments in risk management (market risk, credit risk, operational risk, capital attribution, and liquidity risk), behavioral finance, trading and financial markets innovations, financial engineering, alternative investments and the hedge funds industry, and financial crisis management.

The series will consider a broad range of textbooks, reference works, and handbooks that appeal to academics, practitioners, and students. The inclusion of numerical code and concrete real-world case studies is highly encouraged.

Published Titles

Decision Options®: The Art and Science of Making Decisions, **Gill Eapen**

Emerging Markets: Performance, Analysis, and Innovation, **Greg N. Gregoriou**

Introduction to Financial Models for Management and Planning, **James R. Morris and John P. Daley**

Pension Fund Risk Management: Financial and Actuarial Modeling, **Marco Micocci, Greg N. Gregoriou, and Giovanni Batista Masala**

Stock Market Volatility, **Greg N. Gregoriou**

Portfolio Optimization, **Michael J. Best**

Operational Risk Modelling and Management, **Claudio Franzetti**

Handbook of Solvency for Actuaries and Risk Managers: Theory and Practice, **Arne Sandström**

Quantitative Operational Risk Models, **Catalina Bolancé, Montserrat Guillén, Jim Gustafsson, Jens Perch Nielsen**

Proposals for the series should be submitted to the series editor above or directly to:
CRC Press, Taylor & Francis Group
4th, Floor, Albert House
1-4 Singer Street
London EC2A 4BQ
UK

CHAPMAN & HALL/CRC FINANCE SERIES

Quantitative Operational Risk Models

Catalina Bolancé
Montserrat Guillén
Jim Gustafsson
Jens Perch Nielson

CRC Press
Taylor & Francis Group
Boca Raton London New York

CRC Press is an imprint of the
Taylor & Francis Group, an **informa** business
A CHAPMAN & HALL BOOK

Tables 4.1 to 6.5 and Figures 5.1 to 6.6 are reprinted with permission of the Journal of Operational Risk and the Geneva Papers on Risk and Insurance - Issues and Practice

CRC Press
Taylor & Francis Group
6000 Broken Sound Parkway NW, Suite 300
Boca Raton, FL 33487-2742

First issued in paperback 2022

ISBN 13: 978-1-03-247757-2 (pbk)
ISBN 13: 978-1-4398-9592-4 (hbk)
ISBN 13: 978-0-429-18425-3 (ebk)

DOI: 10.1201/b11602

Publisher's Note
The publisher has gone to great lengths to ensure the quality of this reprint but points out that some imperfections in the original copies may be apparent.

Visit the Taylor & Francis Web site at
http://www.taylorandfrancis.com

and the CRC Press Web site at
http://www.crcpress.com

To our families, friends, and colleagues

About the authors

Catalina Bolancé has been Associate Professor of Quantitative Methods for Economics and Management Science of the Department of Econometrics at University of Barcelona since 2001. She received a PhD in Economics, an M.A. in Marketing, and a B.Sc. in Statistics at the University of Barcelona . She is currently member of the research group Risk in Finance and Insurance and is a specialist in applied nonparametric methods. She has coauthored several undergraduate books on applied statistics that are widely used in Spanish universities and has also published in high-quality scientific journals like *Insurance: Mathematics and Economics*, *Statistics*, and *Astin Bulletin—The Journal of the International Actuarial Association*, the journal of the International Actuarial Association. She has supervised many Master,'s and Ph.D. theses with an outstanding tutoring record. Since 2010 she has participated in a project of the London School of Economics on long-term care insurance sponsored by the AXA research fund. In 2004 she received the insurance international prize awarded by MAPFRE.

Montserrat Guillén has been Chair Professor of Quantitative Methods at the University of Barcelona since 2001 and director of the research group on Risk in Finance and Insurance. She received an M.S. degree in Mathematics and Mathematical Statistics in 1987, and a Ph.D. degree in Economics from the University of Barcelona in 1992. She also received the M.A. degree in Data Analysis from the University of Essex, United Kingdom. She was Visiting Research faculty at the University of Texas at Austin (USA) in 1994. She holds a visiting professor position at the University of Paris II, where she teaches Insurance Econometrics. Her research focuses on actuarial statistics and quantitative risk management. Since 2005 she has been an associate editor for the *Journal of Risk and Insurance*, the official journal of the American Risk and Insurance Association, a senior editor of *Astin Bulletin—The Journal of the International Actuarial Association*, the journal of the International Actuarial Association since 2009, and chief editor of *SORT—Statistics and Operations Research Transactions* since 2006. She was elected Vice President of the European Group of Risk and Insurance Economists at the World Congress of Risk and Insurance Economics in 2010.

Jim Gustafsson has been Head of Nordic Quantitative Advisory Services at Ernst & Young since 2011. He has 10 years of business experience in the insurance and consultancy sector, where he was an employee for several years in the insurance Group RSA with positions such as Actuary, Operational Risk Specialist, Head of Research, and Enterprise Risk Control Director. He received a B.Sc. in Mathematical Statistics in 2004, an M.Sc. in Mathematics in 2005 from Lund University, and a Ph.D. degree in Actuarial Science from the University of Copenhagen in 2009. He is author of over a dozen published articles on actuarial statistics and operational risk quantification, and was awarded for the "Best academic Paper" by *the Operational Risk and Compliance Magazine* in 2007. He was recognized as a "Top 50" Face of Operational Risk by *Operational Risk and Compliance Magazine* in 2009; this award acknowledges the contribution that the recipient has made, and continues to make, to the discipline of operational risk. He is a sought-after speaker in Risk Management and Actuarial conferences and a member of the Editorial Board for the international journal *Insurance Markets and Companies: Analyses and Actuarial Computations*.

Jens Perch Nielsen has been Professor of Actuarial Statistics at Cass Business School in London and CEO of the Denmark based knowledge company Festina Lente. He has a history of combining high academic standards with the immediate practical needs of the insurance industry. Through his company he has managed projects on operational risk, reserving, capital allocation, and risk-adjusted cross-selling methods in non-life-insurance, and he has academic publications in all these areas. In life and pension insurance he has conducted professional work on product development, asset allocation, longevity models and econometric projections and on bread and butter type actuarial day-to-day work. His Ph.D. from UC-Berkeley was in Biostatistics and he is still working on his original topic of general nonparametric smoothing techniques in regression, density and hazard estimation.

Preface

All human activities imply some kind of risk because the possibility of a loss due to accidents or mistakes always exists. Industrial production sometimes requires manipulation of hazardous materials and even the use of sophisticated machinery that is vulnerable to failures. Operational risk is natural in this context and refers to potential costs resulting from errors in normal functioning.

The concept of operational risk has extended from the industrial sector to all other activities, but it is intrinsically difficult to identify. In service provision, some types of disruptions are seldom detected and, moreover, it is very hard to identify what has caused them. So, although operational risk is recognized broadly, little is known about how to handle it.

Insurance is an old method to deal with risk. It is based on the honorable principle of redistributing losses, so that individual exposure can be pooled with others, and the collective faces the expenses and compensates the victim with an economic payment. Nowadays, insurance firms specialize on transferring risks from individual units to portfolios and then to external agents. Insurers also need to keep up to solvency standards as their reputation is essential for a business that is based on the promise to compensate in the event of a loss. The first step toward building an insurance contract is to quantify the risk. But, in fact, there may be errors in the models that are being used for measuring risk, as in many other usual processes that are implemented in insurance activities. Therefore, insurance itself can also be in danger of suffering operational risk.

In the banking sector, as in many other fields, financial transactions are subject to operational errors. So, operational risk is also part of banking supervision.

Given these circumstances, regulators require to measure operational risk as part of the indicators for solvency. This area is quite new and has exploded in the last few years with the existence of the Basel agreements in the banking industry and Solvency II in the insurance sector, which have set the path for international standards of market transparency of financial and insurance service operators.

Measuring operational risk requires the knowledge of the quantitative tools and the comprehension of insurance activities in a very broad sense, both technical and commercial. This book offers a practical perspective that combines statistical analysis and management orientations.

The book provides a guideline to practitioners, going from the basics of what to do with operational risk data to more sophisticated and recent tools that are needed to quantify the capital requirements imposed by operational risk.

This project started when Jens Perch Nielsen was director of research at RSA. A young Swede came into his office in Denmark asking him whether Nielsen would like to supervise his thesis. Two weeks later, Jim Gustafsson was on his way to London to talk to Chief Actuary Dix Roberts and operational risk expert and doctor of philosophy Paul Pritchard. That also led to a close cooperation with Catalina Bolancé and Montserrat Guillén from Spain, experts in how to transform data to improve nonparametric smoothers. Now, a few years and a lot of scientific papers later, Jens, Cati, Jim, and Montse are happy to be able to present this book on operational risk to the academic as well as the practical world.

Dix Roberts and Paul Pritchard have followed the entire process, and they have taken the time to have countless conversations with the authors on "where to go next", which have been invaluable in their search for methods of real practical value. They have also both coauthored papers on operational risk, and the authors of this book are extremely grateful to them for their contribution.

The authors are also grateful to Tine Buch-Kromann who is also a doctor of philosophy from RSA—an expert in transformation herself—who spent energy and time to help supervise Jim's master's thesis. In this way, Jim got the best possible introduction to the transformation approach that has indeed partly been developed by Tine. Tine also coauthored works on operational risk, and she has constantly been an energetic and valuable discussant of our operational risk ideas. They also acknowledge all colleagues from the Riskcenter at the University of Barcelona, who provided many good suggestions and a wonderful atmosphere for doing research with many academic visitors. They thank David Pitt, who read and corrected earlier versions of the first three chapters.

The authors' ambition with this book is to present the highlights of their work on operational risk, and it is therefore not an attempt to cover the substantial field of operational risk. There are important papers, books, and approaches on operational risk out there that give a comprehensive presentation of this topic. The authors have, instead, chosen to cover what they believe they do best: what they see as their own innovative approach to operational risk based on how to get prior knowledge into the estimation. Prior knowledge of operational risk can come in many packages, one of the most important being prior knowledge from external data, related to, but different from the data under investigation. Other prior knowledge could be on parametric shapes that have shown themselves to be useful in other similar studies, or even nonparametric shapes taken from sources other than the data at hand.

The book includes real-life examples and discusses the results, showing the usefulness of the methods described in the different chapters. It also provides an overview of the main difficulties arising in practice when quantifying operational risk in an insurance company or in a bank. There are many special features and sources of error that do not occur in other fields as they do in banks and insurance companies.

In operational risk, one usually has to combine internal data and external data (i.e., information arising from the company with information arising from other firms); this topic has not been discussed in other books. The question of underreporting (not knowing the total frequency and severity of all errors that occur in the normal function of the company) is also one of the contents that is included in this book. Underreporting is not extensively mentioned in the existing literature.

The last chapter of this book is a self-contained guided example with programs in R and SAS©, which can be useful to practitioners in order to implement most of the methods that are presented along the text. Data to reproduce the results in Chapter 7 are available from the authors.

Everything the authors have done in this book has been developed in the spirit that it should be practical, relevant, and based on academic standards. It is now up to the reader to judge whether they have been able to live up to their own ideal.

Contents

List of Figures

List of Tables

Chapter 1

Understanding Operational Risk

1.1 Introduction

Operational risk refers to the possibility of unexpected events that occur as a consequence of alterations in normal functioning. One would expect that stopping the operations would definitely eliminate the possibility of any error, so operational risk includes all things that can happen in day-to-day activity. For most of us, it is easier to imagine what operational risk is when we think of industrial production, where machinery is being used. In that case, equipment may get broken and therefore the possibility exists that fabrication has to be stopped. A machine that cannot work is something one can physically inspect, and there is no doubt that it has caused a stop in the usual assembly chain. One can see this example as the simplest one in operational risk measurement: one can count an average number of hazards per unit produced (or per thousand units produced) and get an average estimate of the probability of operational failure. This is what engineers have called quality process management.

In the financial and insurance sector, the central activity has to do with contracts, investment, economic transactions, and customer relationships, but the risk of any of these activities not functioning correctly still exists as in industrial plants. Events without physical consequences are more difficult to detect. A gross failure for an insurance company would be, for example, a technical actuarial mistake in the design of a new product. What makes this example of operational malfunction especially unique? First, the effects of the mistake may become evident long after the product has been marketed. Second, the negligence can lead the firm to severe losses and, if discovered by regulatory bodies, can imply huge penalties. Third, and most important in this sector, the miscalculation does not imply an immediate stop to the activities, unlike a machine broken in the middle of a production chain. So, managers can become aware of the problem too late, only once the economic consequences are evident.

When we want to quantify operational risk, the most difficult part is to identify what can go wrong. Besides, one should calculate how often this can happen and what the consequences of its occurrence are. Once the phenomenon has been characterized, then statistical methods will allow extrapolations. This book is about straightforward methods that would allow the practitioner to go one step further from observed frequencies and averages. One could, moreover, be concerned about what would be a bound for the losses derived from operational risk with a 99% probability. This requires examining the possibility of occurrence outside the range of regular occurrences, and this is what the theory of extremal events helps to address.

1.2 Our Approach to Operational Risk Quantification

The quantification of operational risk meets a number of difficulties because of sparse data. Normally, there are not many collected internal operational risk observations for one given company since legislative requirements are very recent and there is a systematic underreporting of operational failures. Besides this, companies are really interested in events driving their capital, the most expensive losses they could be exposed to, even if they are very infrequent. These are called *tail events* because, in statistical terms, they represent the tail behavior of the operational risk distribution. It is really tail events that cause companies to fail.

Therefore, sparse data, underreporting, and extreme values are three reasons for operational risk being extremely difficult to quantify. Based on market experience, one cannot just do a standard statistical analysis of available data and hand results in to the supervisory authorities. It is necessary to add extra information. In this book we call extra information "prior knowledge." While this term originates from Bayesian statistics, we use it in a broader sense: prior knowledge is any extra information that helps in increasing the accuracy of estimating our internal data. In this book we consider three types of prior knowledge, and one is the prior knowledge coming from external data. First, one has data that originate from other companies or even other branches (for instance, insurance companies have so few observations that they need to borrow information from banks) that are related to the internal operational risk data of interest—but only related; they are not internal data and should not be treated as such. Second, one can get a good idea of the underreporting pattern by asking experts a set of relevant questions that in the end make it possible to quantify the underreporting pattern. Third, we take advantage of prior knowledge of relevant parametric models, and it is clear that if you have some good reason to assume some particular parametric model of your operational risk distribution, then this makes the estimation of this distribution a lot easier. Prior knowledge on a parametric assumption can be built up from experience with fitting parametric models to many related data sets, and in the end one can get some confidence that one distribution seems more likely to fit than another distribution.

For our studies, we have so far come up with the prior knowledge that often the generalized Champernowne distribution is a good candidate for operational risk modeling. The book also considers the art of using this prior knowledge exactly to the extent it deserves. This is, when we have very little internal data at our disposal, we have to rely heavily on prior information.

When data become more abundant, the internal data set at hand takes over, downplaying the role of prior information. This is pretty much the basic idea of the celebrated credibility theory of non-life-insurance; however, in this book we show that a similar flexible approach results from the more transparent transformation approach developed throughout the book. The underlying idea of the transformation approach is the following: if your prior knowledge is correct, then you can transform your data accordingly so that you end up with simple, independent, identically uniformly distributed stochastic variables. So, perform this transformation out and check whether the transformed stochastic variables indeed look to be uniformly distributed, or rather, whether we have enough data to properly verify how the transformed data differ from simply uniformly distributed variables. Transforming to uniform variables means that we transform into a bounded interval. This is exactly the same approach suggested in copula transformation in the multivariate case. Another advantage is the flat symmetric shape of uniform distributions. When transforming into the unit interval, an additional advantage is that values can be interpreted as transformations of quantiles of the original data.

Our approach based on kernel smoothing has the appealing property that small data sets imply that a lot of smoothing is necessary, and this again implies that we end up not differing too much from our prior knowledge. However, when data becomes more abundant, less smoothing is optimal, which implies that we have sufficient data to reveal the individual distributional properties of our internal data downplaying the importance of prior knowledge. Another key result of this book—based on real data—is that our transformation approach also has a robustifying effect when prior knowledge is uncertain. We show, for example, in a real data study where we use three very different parametric models with very different conclusions of the implied operational risk that, after our transformation approach, we get similar results independent or almost independent of the parametric assumption used as prior knowledge. Therefore, the message of this book is

Find as much prior knowledge as possible and then adjust this prior knowledge with the real observations according to how many real observations are available.

1.3 Regulatory Framework

At the beginning, operational risk was defined as all sources of risks that could arise other than market and credit risk. The Basel Committee on Banking Supervision had already made an explicit reference to operational risk in 1998 and, since then, it has become part of the capital adequacy requirements in financial institutions. Current practice is based on the Basel II accord, while Basel III is already on its way.

In Basel II, banks are allowed to use three different approaches to operational risk quantification. The simplest one is called the basic indicator, and it is just a measure of risk equal to a constant percentage (15%) of an average gross income over the previous three years. The second approach is called the standardized approach, which goes a bit further in the sense that the bank's activity is divided into eight business lines. These eight lines are generally accepted for all banks and include corporate finance, trading and sales, retail banking, commercial banking, payment and settlement, agency services, asset management, and retail brokerage. The risk measure is then a sum of a fixed percentage of average gross income for each line. The percentage used varies between lines of business, ranging from 12% to 18%. The assumption on both the basic indicator approach and the standardized approach is that there exists a linear relationship between volume (gross income) and operational losses, but there is absolutely no consideration for risk control or risk mitigation. In fact, banks or financial institutions with a similar gross income would require the same amount of capital, no matter to what extent they may differ in the implementation of internal controls, risk management standards, audits, or quality of processes.

The third approach, and most sophisticated, in the Basel II framework is called AMA, which stands for Advanced Measurement Approach. When using this method, banks are allowed to develop internal models that capture the characteristics of their own situation, taking into account risk mitigation controls and idiosyncrasies.

This book contributes to Basel II and to the AMA, which also seems to be part of Solvency II.[1] In Solvency II, insurance companies are allowed to lower their solvency requirement if they can provide internal high-quality models that convincingly show that the standard formula methods are overly prudent. Solvency II constitutes a fundamental review of the capital adequacy regime for the European insurance industry and aims at establishing a revised set of EU-wide capital requirements and risk management standards that will replace the requirements of the present European regulatory regime. Solvency II is based on a three-pillar system that corresponds to Basel II. Pillar I concerns capital requirements, the Minimum Capital Requirements (MCR) being smaller than the Solvency Capital Requirements (SCR). SCR is the capital requirement of the insurance company. The regulating authority (for instance, FSA, which is the Financial Services Authority in the UK) increases its control when capital gets below the SCR. When capital gets below the MCR, FSA can take over the management of the insurance company and perhaps even stop the underwriting of new business.

Capital requirements are calculated based on total risk quantification in the insurance company, which consists of insurance risk, credit risk, market risk, and operational risk. While Solvency II specifies a standard formula with predefined correlations for the two capital requirements, insurance companies are

[1] Solvency II is the European regulation structure that specifies how insurance companies must manage their risks and reserve prudent solvency margins.

allowed to calculate the economic capital with internal statistical models if they show that this is prudent. This provides an incentive to insurance companies to spend money and energy on internal models

1.4 The Fundamentals of Calculating Operational Risk Capital

According to the AMA, institutions must have sound estimates of all quantiles up to 99.9%. The institution must furthermore maintain rigorous procedures for developing operational risk models and validating the model. Basel II specifies guidelines and recommendations for the use of external loss data, which underline the importance of using relevant external data, especially when there is reason to believe that the institution is exposed to infrequent yet potentially severe losses. The guidelines are first and foremost concerned with documentation of the events that led to the losses. For example, Basel II requires information on actual loss amounts, on the scale of business operations where the event occurred, on the causes and circumstances of the loss events, and other information that could help in assessing the relevance of the loss event for other institutions; similar demands are made for the treatment of internal loss data. Another demand in Basel II is that it must be easy to place the internal data set in the relevant supervisory categories and that the internal and external data set must be provided to the supervisors upon request. Similar guidelines will certainly be specified in Solvency II.

A good way to efficiently compute operational risk measures is to combine two recent developments:

- A statistical analysis of the quantitative impact of the failure to report all operational risk claims.
- Recent development of smoothing methods that are capable of estimating nonparametric distributions with heavy tails.

These two developments are described in detail in the following chapters.

When estimating operational risk losses, it is a major obstacle that not all losses are observed. To estimate such an underreporting function from the data itself is an incredibly complicated mathematical deconvolution problem, and the rate of convergence of the deconvoluted estimators is often very poor. On top of that, the deconvoluted estimators often rely too heavily on the underlying assumptions about the underreporting function. Inspired by a famous statistician David A. Freedman, we therefore decided to "put our shoes on" (Freedman's phrase) and go out in the world and collect the crucial data that we need in order to improve our estimation. After extensive interviews with experts in the insurance market with a long experience in a major non-life-insurance company, and after a qualitative decision process, we deduced our best guess of an underreporting function.

We have seen that an underreporting function created in collaboration with experts and managers simplifies the theoretical problems and yields a solution that is closely related to what we would have obtained if we had observed all the losses without underreporting.

In this book we use a nonparametric smoothing technique to estimate the distribution of operational losses when underreporting is taken into account, and later, mixtures with external data will also be considered.

An example will be given on how the method is applied to a database of operational risk from a financial institution with six major lines of business. In the database, there are sufficient data in each business line so that the well-known credibility approach is not necessary. However, it is possible to combine a credibility technique with our nonparametric smoothing method in order to extend the method to more sparse data sets. The method will be applied to the six lines of business and combine an internally estimated frequency of expected reported claims with an externally estimated distribution of operational risk losses. The externally available database of operational risk is from financial institution because there are not yet many reliable data on operational risk in the insurance industry. Insurance companies are therefore forced to use operational risk data from other financial institutions.

Many financial services organizations are now utilizing loss data for the purposes of calculating operational risk capital requirements, potentially arising from either regulatory requirements or indeed from a desire to integrate capital-sensitive management within their organizations. Instinctively, the use of internal loss experience directly or as a means of deriving distribution parameters from which simulations can be made is most appealing. However, several factors mitigate against its effectiveness when considered alone: first, the data are backward looking based on historical events; the company profile may have changed, and should any large losses have occurred, it is likely that controls will have been improved to prevent a reoccurrence. A greater problem nonetheless is that the regularly encountered losses may provide limited information on the size and frequency of large, rarely occurring losses that are the major factor in determining capital requirements. With this in mind, organizations have recognized the value of obtaining loss data from outside their company either through data sharing consortia or through publicly reported losses.

1.5 Notation and Definitions

The calculation of operational risk capital under the advanced measurement approach requires that a model is specified. In fact, it becomes natural and intuitive to count the occurrence of events and to consider the size of each loss. This ends up being what is known as a frequency and severity model.

The notation used throughout this book is as follows. We will denote random variables by capital letters; generally, X will be used for severities, and N for frequencies. Data in operational risk are often a combination of two random sources. On the one hand, we have the random number of events that occur due to operational failures, which we call the frequencies and which are discrete and nonnegative. On the other hand, we have the severity of the loss due to the occurrence of the operational event. The loss is assumed to be continuous and nonnegative. Therefore, in our presentation, most of the continuous random variables that model the severity are assumed to be nonnegative and heavy-tailed, as this is the common feature in our operational risk context.

For a continuous random variable X, we denote by $F(\cdot)$ the cumulative distribution function (cdf), which is defined by

$$F(x) = \Pr(X \leq x).$$

The cdf satisfies three properties:

- $\lim_{x \to \infty} F(x) = 1$ and $\lim_{x \to -\infty} F(x) = 0$.
- $F(\cdot)$ is right continuous.
- $F(\cdot)$ is nondecreasing monotonous.

If it exists, the derivative of $F(\cdot)$ is called the probability density function (pdf) denoted by $f(\cdot)$, and then,

$$F(x) = \int_{-\infty}^{x} f(t)dt.$$

For a discrete random variable N, a similar notation will be used. When it is necessary, a set of random variables will be required, and then subscript i will generally be used and the sample size will be denoted by n.

A *risk measure* for random variable X will generally be denoted by $\rho(X)$. In calculating capital requirements, the most widely used risk measure is Value-at-Risk at the α-level, $VaR_\alpha(X)$. This is known in statistics as the quantile function, which is defined as

$$\begin{aligned} VaR_\alpha(X) &= \inf\{x \in R : \Pr(X > x) \leq (1-\alpha)\} = \\ &= \inf\{x \in R : F(x) \geq \alpha\}, \end{aligned}$$

where $\alpha \in (0,1)$.

The meaning of Value-at-Risk is intuitive, and it indicates that the proportion of values of X that are observed over the threshold $VaR_\alpha(X)$ is $(1-\alpha)$, that is, $100(1-\alpha)$ in 100. If $\alpha = 99\%$, then a value of X above $VaR_\alpha(X)$ should be observed with probability 1%.

There are simple and straightforward expressions for $VaR_\alpha(X)$ when X is assumed normally distributed or t-Student with ν degrees of freedom. Otherwise, more sophisticated, numerical methods need to be used to approximate this risk measure.

Another possibility, which is used in the Swiss Solvency Test approach to risk measurement in insurance companies, is known as Tail-Value-at-Risk at the α-level and denoted by $TVaR_\alpha(X)$. It is also referred to as Expected

Shortfall ($ES_\alpha(X)$), sometimes with slight definition changes due to how continuity and discontinuity in $F(\cdot)$ is assumed. The definition of $TVaR_\alpha(X)$ for a continuous random variable X is

$$TVaR_\alpha(X) = \frac{1}{1-\alpha}\int_\alpha^1 VaR_u(X)du.$$

Since Tail-Value-at-Risk can be expressed as a conditional expectation of random variable X, it is also known as Conditional Tail Expectation ($CTE_\alpha(X)$):

$$TVaR_\alpha(X) = CTE_\alpha(X) = E(X|X \geq VaR_\alpha(X)).$$

Current regulation principles that rely on risk modeling approaches are based on the previous risk measures.

1.6 The Calculation of Operational Risk Capital in Practice

Capital requirement is calculated as a simple function of the estimated risk measure, that is a proportion. In order to achieve a good internal model, the main challenge is to find an adequate statistical model for the aggregated loss. This is known as Loss Distribution Analysis (LDA). There are three basic methods to effectively calculate the risk measure in each case. The variance and covariance method requires the first and second moments to be estimated and a parametric distribution to be assumed. Needles to say that choosing the parametric distribution is a rather controversial hypothesis. Historical simulation can only be implemented when the amount of data is substantial. In this case, the empirical distribution is often used so that no parametric assumption is needed. Finally, the Monte Carlo method is used, where the distribution of total loss is obtained by generating random values corresponding to losses. This is statistical simulation, but in this case, a model is assumed for the stochastic behavior of losses, and the dependence structure must also be fixed.

By letting insurance companies be able to develop internal models for operational risk and by using these to determine the amount of appropriate capital, the idea is that this should lead to a better treatment of a specific risk. Insurance companies that choose not to develop internal models will be forced to base their capital amount on regulatory standards, which are formed for a universal application and may not be optimal for a specific company. The operational risk number under the standard formula is suggested to be a percentage of other risks in the company. It can be interpreted that the operational risk number is driven by premium and reserve volatility, which, of course, is not fully justified. If a company decides to take a more sophisticated path, the internal models will be subject to regulatory supervision and approval. The supervision will not only be made on the adequacy of sophistication and appropriateness of the internal models but also on the overall monitoring and management of operational risk. This forms a regulatory reason for a more accurate modeling of operational risk.

An insurance company must have a sufficient supply of risk based capital in order to cover possible operational losses. To this end, the most important issue for a company is to have a clear and correct understanding of operational risk. This risk has always existed and is at the core of every organization. However, businesses start to develop a recognition of operational risk as a stand-alone risk, which has to be measured, monitored, and managed on its own and in conjunction with the other risks.

Since operational risk capital assessment is new for most insurance companies, the level of model sophistication is quite low. This, of course, is a result of many things such as lack of reliable data or manpower, or even of technical experts in the operational risk management team. Therefore, the calculation of operational risk capital has tended to follow a very simple approach for insurance companies. A company should not just choose the most advanced methodology but pick one that makes operational risk management workable.

When a company is able to measure and allocate the appropriate amount of capital with models that make them comfortable, the business and the individual business units would be able to take risk-based decisions both on a daily basis and with a more long-term strategic view. This is also a requirement having an internal model for operational risk in place. A good treatment of operational risk has big effects on the company's competitiveness. This implies getting a competitive advantage in the way the company sets prices, makes strategic planning, and runs the business on a daily basis.

In general, insurance companies use a hybrid approach to calculate operational risk capital, with scenario analysis as its primary measure. The approach incorporates aspects of industry basic, standard, and enhanced models used elsewhere in the financial services industry. It uses three separate techniques to triangulate a result, where the LDA and Benchmarking approach validate the results derived from the Scenario Analysis. The reliance on Scenario Analysis is because it is deemed to be a more robust method, clearly enabling senior management to understand the capital and risk management impacts of good risk management. The quantification process to calculate the capital figure for insurance companies follows normally then as a step- by-step procedure. To obtain the hybrid approach, the first method that a company starts to develop is the Scenario Analysis approach. This is because the company has not started to collect data. Thereafter, the company needs to validate this model, and two other methods are in general undertaken—the loss distribution analysis on external data and industrial benchmarks. These provide confidence as to the accuracy and robustness of the primary measure.

The scenario analysis approach estimates the probability and size of loss for given confidence levels and is based on inputs from experts. Normally, this is achieved by internal workshops to complete a scheme of operational risk scenarios. In each workshop, a number of subject matter experts are asked to formulate the supposed types of losses that could occur under a specific risk—and give their estimates of the typical loss amount—one for 20-year loss and other for the annual loss frequency for each of the risks the company has defined to cover operational risk (e.g., the Basel II categories). With this information, a spreadsheet is created that forms the basis of the capital calculation that should be held for operational risk. To be able to know the regulatory and internal percentiles, the subject matter experts' knowledge is used for distribution fitting, and then the adjusted distribution of losses is obtained using Monte Carlo simulation.

When mixing information for a specific event risk category, two severities are given for specific percentiles. Then, by assuming a random variable X that follows a two-parameter distribution F, for example the Weibull distribution, one could obtain the estimated parameters $\theta = \{\theta_1, \theta_2\}$ for this event risk category. One needs to solve a system of equations such as

$$\begin{cases} VaR_{q_1} = x_{q_1} \\ VaR_{q_2} = x_{q_2} \end{cases}$$

Here, expert opinion is used on both the severities $\{x_{q_1}, x_{q_2}\}$ and the corresponding percentiles $\{q_1, q_2\}$.

For the frequency, the most common assumption in the industry is that the losses arrive as a Poisson process. Also here, the intensity is provided by expert opinion since the intensity is estimated by the observed losses (expert opinion annual frequency) divided the duration (expert opinion 1 year).

With this information, a total loss distribution could be estimated by combining the severity and frequency distribution. This could be done on individual event risk categories, or one could have a total one for the whole company.

The workflow process by using this method is straightforward: data analysis, model assumptions, risk quantification, and capital requirement approximation. Once the first approximation is ready, the next step would be to check whether the resulting economic capital requirement is less than the one that would be obtained under the basic model or the standardized model, although perhaps a sophisticated internal model that suggests higher levels of capital being held than the regulator standard is very useful information for the company. Prudent risk management may dictate that the company should hold capital reserves at a higher level than required by the regulator as a result of its own internal investigations.

1.7 Organization of the Book

The book is organized as follows. Chapter 2 is devoted to the presentation of classical parametric severity distributions, and illustrations using a real data set are provided. Chapter 3 is dedicated to the estimation of the severity distribution under the semiparametric perspective. This allows much more realistic results on how both small and large losses are captured under the same distribution. Chapter 4 describes how to incorporate both internal and external information in the same model. Chapter 5 deals with the underreporting problem and describes the fundamental challenges and the development of a model that capture these elements. Chapter 6 combines all previous methods in one comprehensive approach to operational risk modeling that can be implemented in practice with minimum restrictive assumptions. Finally, Chapter 7 presents the practical implementation that can be used as a guide to quantify operational risk.

Throughout the book, several sets of data have been used in order to discuss examples and interpret results. Data used in Chapter 7 have been generated for illustrative purposes. Programs included in Chapter 7 cover the methods described in Chapters 2 to 5. The methods in Chapter six can easily be implemented as a combination of the previous ones.

1.8 Further Reading and Bibliographic Notes

There are several books and articles related to operational risk, but none has the same focus as this one. Cruz's [22] was one of the pioneering monographs on this topic and addresses how to model, measure, and hedge operational risk. Panjer's book [62] has an excellent presentation of the topic and is an essential reference for those seeking the statistical foundations underlying most common methods. Some other general and valuable references are the books by Chernobai et al. [17], Davis [24], Alexander [1], and King [55], and an interesting contribution on the invention of operational risk in the beginning of the 90s is Power [63]. de Fountnouvelle et al. [26] indicates the difficulty of creating good operational risk data sets.

McNeil et al. [60] is an excellent reference on risk management. It also covers operational risk, especially from the banking perspective. Note also the introduction to and overview of operational risk modeling according to the Basel II legal documents and summary on observed practices and issues as well as suggested approaches for measuring and quantifying operational risk given by Embrechts and Hofert [30].

A valuable reference for operational risk modeling and management from a very broad perspective is Franzetti [37]. That book covers a wide range of problems related to operational risk and is extremely useful for understanding the global problem; however, the tails of the distribution of event severities are dealt with using standard parametric models. Apart from being a recommendable monograph for managers, readers should refer to it for, among other parts, a very interesting chapter on audit, reporting, and disclosure of operational risk.

More relevant contributions on quantitative models for operational risk can be found in [16], [27], [32], [61], [33], and in many references that are provided in the following chapters.

Chapter 2

Operational Risk Data and Parametric Models

2.1 Introduction

The importance of sound operational risk management is increasing in the insurance sector, particularly since the introduction of the Individual Capital Adequacy Standards regime in the UK (CP190) and the upcoming EU Solvency II Directive. These Directives, compared to Solvency I, highlight a sound risk framework, and operational risk is one key risk module in the new Solvency II regime.

Reliable historical loss data is essential to operational risk management and modeling of risk exposures, and gathering good-quality information on operational risk losses in the insurance industry is a challenge, perhaps an even bigger challenge than the one banks face.

Insurance companies have not collected internal data for that many years; therefore, a deeper view of the internal databases will result in very scarce samples. Insufficient data is the result of several circumstances. For instance, insurers may have not been exposed to large losses, parts in the organization may have not reported operational risk losses as one would expect, or the losses they suffer may have been ignored or hidden.

There is no doubt that the general view in the insurance industry is that historical loss data are essential for the effective measurement and management of operational risks, and since operational risk losses in insurance are not reported on a continuous basis, the internal losses are thus unlikely to provide a full picture of the universe of risks faced by the organization. The internal losses could then be seen as a biased sample of the set of potential losses because they reflect idiosyncratic features such as the company business, risk appetite, organizational culture, and internal control environment.

Information on losses experienced by other companies can fill some important gaps according to internal samples, but to deliver meaningful results and thereby use these data points in practice, it is vital that these losses are comparable to the losses the company might experience. It is necessary to find robust methodologies for scaling the size and the number of external operational risk losses, thereby adjusting for potential biases that affect external databases and make them comparable with the internal sample. Adjusting for potential scaling biases is particularly important when external and internal losses are pooled for operational risk management and economic capital calculation.

This chapter examines an example of internal and external data that are used throughout some parts of this book. Thereafter we continue by discussing several probability distributions of the loss severity. This analysis involves continuous statistical distributions. Having a good understanding of amounts of losses is important not only for modeling purposes but also for identifying effective management and mitigating actions.

2.2 Internal Data and External Data

The lack of data on actual losses is one of the main stumbling blocks when modeling operational risk exposures. This issue has already been faced in the banking sector but is likely to have an even bigger impact on the insurance industry since operations involve fewer transactions and less trading, which are known as important drivers of operational failures. An internal operational loss database is a key input in any effective risk management framework. However, internal data on operational risk losses are often limited and potentially biased, not only due to internal collection problems but also because large operational losses in insurance do not happen very often and take time to crystallize.

Internal historical loss events reflect the specific experience and history of the organization, normally limited years, and are therefore unlikely to provide a complete picture of the universe of major operational risks that the company faces. Sources of information on loss events that occurred in comparable companies can sometimes be helpful to fill the gaps. A natural way to overcome the lack of data issue is to merge internal data with external loss data for modeling purposes. However, this data aggregation exercise is a challenge because companies have different business profiles, sizes, and controls that are likely to affect the size and frequency of their operational losses. Failing to adjust for these factors when mixing loss data may lead to flawed estimates and misleading conclusions. There exist several different sources of external data in the industry, but none of these is easy to integrate in the models directly.

The most well known data source is the general information called *publicly available losses*. These databases contain operational risk losses that, because of their magnitude, filter through to the public domain. Information about incidents is obtained from the media such as news reports, specialized publications, and so on. Therefore, these loss events tend to be linked to low-frequency, high-impact operational failures and are normally used to capture the tail.

Another data source is the *commercial database using proprietary loss data*. These databases contain information on operational risk loss events experienced by financial institutions that are not necessarily in the public domain. Some of the incidents are identified to insurable losses, or the data may have

Table 2.1: *Statistical summary of the internal and external data*

Data	n	Mean	Median	Std. Deviation	Maximum
Internal Data	104	0.054	0.014	0.134	1.053
External Data	831	0.287	0.012	1.920	31.400

Note: Size is expressed in monetary units.

been extracted from insurance claims files, whether the claim was made by the company or not.

The third source, and probably the one with less peculiarities, is the *consortium-based loss data*. These databases comprise loss events reported to a consortium by its members, who in return get access to anonymized, pooled industry data on operational risk loss events and near-miss incidents. The members of the industry consortium pay a subscription fee and voluntarily commit themselves to feeding the database with their individual internal loss events so long as confidentiality of key information is protected.

All external databases bring a threshold problem according to the minimum collection threshold. The threshold above which public losses are recorded is likely to be much higher than the reporting threshold for consortium data.

Throughout most chapters of this book, the same event risk category, *Business disruption and system failures*, will be analyzed and utilized in different model setups. We make use of an internal sample and one external sample, and for the latter, a filtering process has been applied rigorously. For example, certain losses may be associated with a particular business activity, geography, or even size, and there has been a scaling process to reflect the control standard in the organization before using the external data. Table 2.1 shows some statistical characteristics owing to the two data samples, and Figure 2.1 presents a visual view of the two samples. Here the x-axis presents the sequence number of losses, and the y-axis indicates the size of the losses.

In Figure 2.1, we note that the size of the losses is very different in the internal and the external samples. The y-axis for internal losses ranges from 0.0 to 1.1, whereas the y-axis for the external data plot ranges from 0.0 to about 35.0. The reason for the different scales is that data from the external sample have been obtained from companies that may be larger than the company from which internal data were obtained. The visual effect is that both data sets have a similar coefficient of variation, so this means that a rescaling procedure would solve the problem of combining both data sets. We will investigate mixing and rescaling in the following chapters.

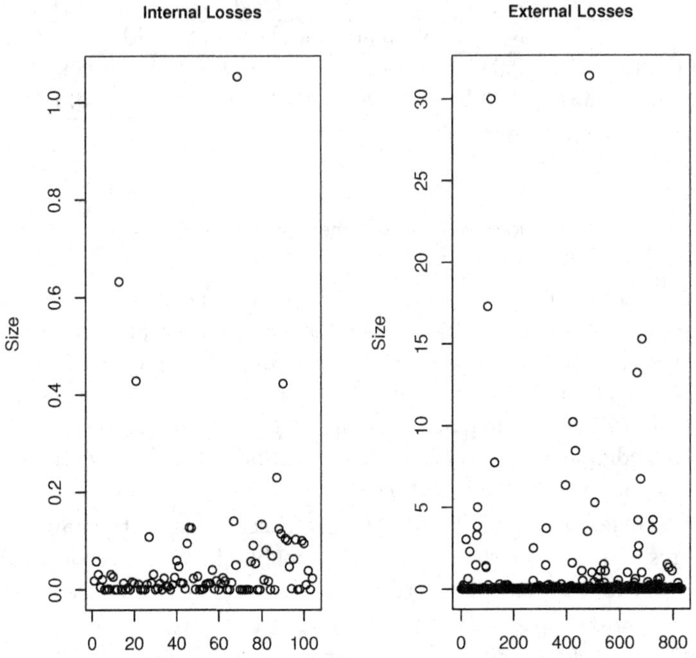

Figure 2.1 *Plotted internal and external operational risk data originating from the event risk category Business disruption and system failures.*

2.3 Basic Parametric Severity Distributions

Financial institutions, such as insurance companies and banks, must have a sufficient supply of risk-based capital in order to cover possible operational losses. To obtain this, one of the most important issues for a financial institution is to have a clear and correct understanding of this particular risk, which corresponds to the uncertainty of future outcomes of a current decision or situation.

We can measure risk because different outcomes occur with different probabilities and the field of all possible outcomes, and their corresponding probability of occurrence could be explained by probability distributions. Because of this, analysts at financial institutions are highly interested in developing an appropriate loss distribution for these outcomes. Loss Distribution Analysis (LDA) is a method used to calculate the necessary amount of risk-based capital. The LDA approach involves modeling the loss severity and the loss frequency separately and then combining these via a Monte Carlo simulation or other statistical techniques to form an aggregated loss distribution.

Modeling the severity of losses usually involves analysis of parametric distributions such as the lognormal, Weibull, or Pareto distributions. These are distributions that are common in the financial sector. However, it is very difficult to find a parametric distribution that fits all sizes of severities: (1) low severity—which occur frequently, (2) high severity—which occur sporadically, and (3) catastrophic severity—which occur a few times per decade. However, modeling any event in nature involves the risk of introducing modeling errors, for example, when using a model to solve a simple problem. This would not only mean a waste of resources but also a lack of understanding on the part of the practitioner when using the model in his or her daily work.

This section explores several possibilities to capture the severity losses for operational risk assessment. The severity distributions describes the size of the losses. Modeling size is rather complicated in comparison to loss frequency modeling due to the unpredictable range of high-severity events. The main issue is to choose a distribution that can cover all losses. Our main interest in this section is analyzing several parametric distributions on the two data samples presented in the previous section.

In total, we introduce five different parametric distributions in this section, and we begin with the most light tailed. As we proceed in this section, the tail will become heavier. We will not clarify which severity distribution is optimal for the two samples that are used for illustration. In the next section the generalized Champernowne distribution will also be presented.

The cdf (cumulative distribution function) and pdf (probability distribution function) for the basic parametric distributions for severity considered here are presented in Table 2.2.

We have estimated the basic densities using maximum likelihood in both the internal and external sample. Some selected results are shown in Figures 2.2 to 2.4. We do not represent all the densities because some characteristics are redundant. So, only the most representative are displayed in order to point out the most interesting features.

In the following figures, we present the parametric pdf estimate and the original data, which appear as dots along the x-axis. The upper part corresponds to the internal data set and the lower part shows the external data set results. On the left, the plots show the lower values of the observed sample, and on the right the larger observed values are presented. Low and high values are divided in order to see the parts of the densities where data are scarce. The scale of the figure is the same in the upper (internal) and lower (external) plots for easier comparison. No internal data above size 1.1 have been observed, and so, there are no plotted dots on the x-axis for the internal data figures for

Table 2.2 *Densities (pdf) and cumulative probability functions (cdf) for some selected distributions*

	pdf	cdf	Parameters	Domain
Logistic	$\frac{e^{-(x-\mu)/\sigma}}{\sigma\left(1+e^{-(x-\mu)/\sigma}\right)^2}$	$\frac{1}{1+e^{-(x-\mu)/\sigma}}$	$\mu \in R$ $\sigma > 0$	$x \in R$
Exponential	$\lambda e^{-\lambda x}$	$1-\lambda e^{-\lambda x}$	$\lambda > 0$	$x \geq 0$
Gamma	$\frac{1}{s^a\Gamma(a)}\left(x^{(a-1)}e^{-x/s}\right)$	$\frac{\gamma(a,x/s)}{\Gamma(a)}$	$a,s > 0$	$x > 0$
Weibull	$1-e^{-(x/b)^a}$	$\frac{a}{b}\left(\frac{x}{b}\right)^{(a-1)}e^{-(\frac{x}{b})^a}$	$a,b > 0$	$x \geq 0$
Lognormal	$\frac{1}{x\sigma\sqrt{2\pi}}e^{-\frac{1}{2}\left(\frac{\log(x)-\mu}{\sigma}\right)^2}$	$\Phi\left(\frac{\log x-\mu}{\sigma}\right)$	$\mu \in R$ $\sigma > 0$	$x > 0$

Notes: $\Phi(\cdot)$ is the cumulative distribution function of a standard normal.
$\gamma(a,x/s)$ is the lower incomplete gamma function.

large values (upper right corner). Nevertheless, some densities may have mass in that domain area as an extrapolation from the lower part of the density. Only parametric density estimates are presented in this chapter.

Figure 2.2 presents the results for the logistic distribution. Logistic distribution is the lightest tailed distribution among the ones used in this book. The maximum likelihood estimates for the parameters of the logistic distribution that is assumed for the external data are unable to produce a shape that represents correctly the density for data above size 5. The density is rapidly decreasing toward zero as x tends to infinity, but the tail becomes too light. The plot on the right lower side shown in Figure 2.2, represents observed data in this domain, while the density is already zero. So, for the external data sample, logistic distributions fails to capture observations of size above 5. Similar behavior occurs when using exponential distribution.

Figure 2.3 shows the density fit for the Weibull distribution for the same two sample data. The right-hand-side plots in this figure show that the right-hand-side tail of the density is heavier than the one obtained for the logistic distribution, and one can already see a light tail density above zero for loss values exceeding size 5. However, for external data, the right tail is still not heavy enough to capture some large observations, that is, there are two observed losses of size around 30, and the Weibull density estimate is almost zero in that area. Similar results are obtained for the Gamma distribution.

Figure 2.4 shows the maximum likelihood estimation of the lognormal densities for the internal and external data. The right-hand-side tail is now heavier than the one for the logistic and Weibull distributions so that the plots on the

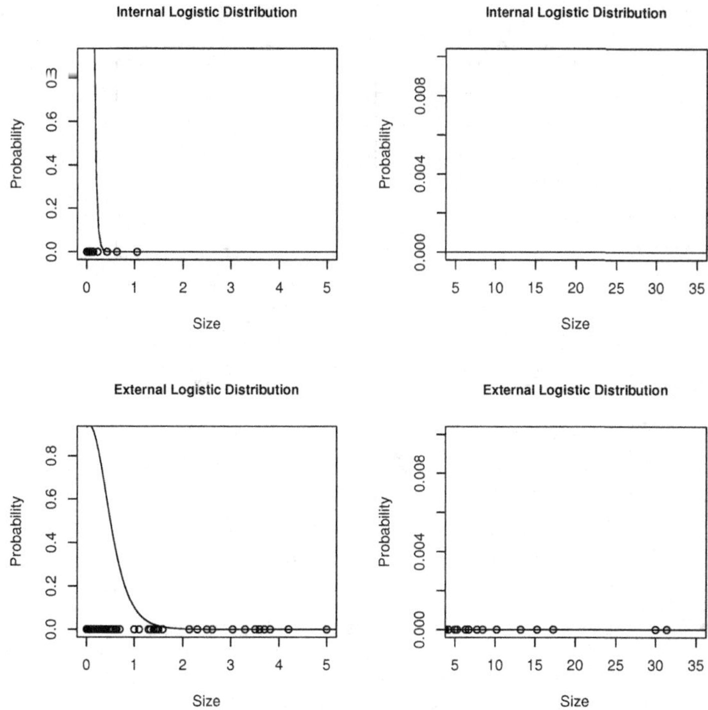

Figure 2.2 *The estimated density for internal losses (above) and external losses (below) operational risk data using the logistic distribution.*

right-hand-side present a strictly positive density curve. The density is still too low for large losses.

2.4 The Generalized Champernowne Distribution

The Champernowne distribution was mentioned for the first time in 1936 by D.G. Champernowne when he spoke on "The Theory of Income Distribution" at the Oxford Meeting of the Econometric Society. Later, he gave more details on the distribution and its application to economics. The original Champernowne distribution has density

$$f(x) = \frac{c_*}{x\left(\frac{1}{2}\left(\frac{x}{M}\right)^{-\alpha} + \lambda + \frac{1}{2}\left(\frac{x}{M}\right)^{\alpha}\right)} \qquad x \geq 0, \tag{2.1}$$

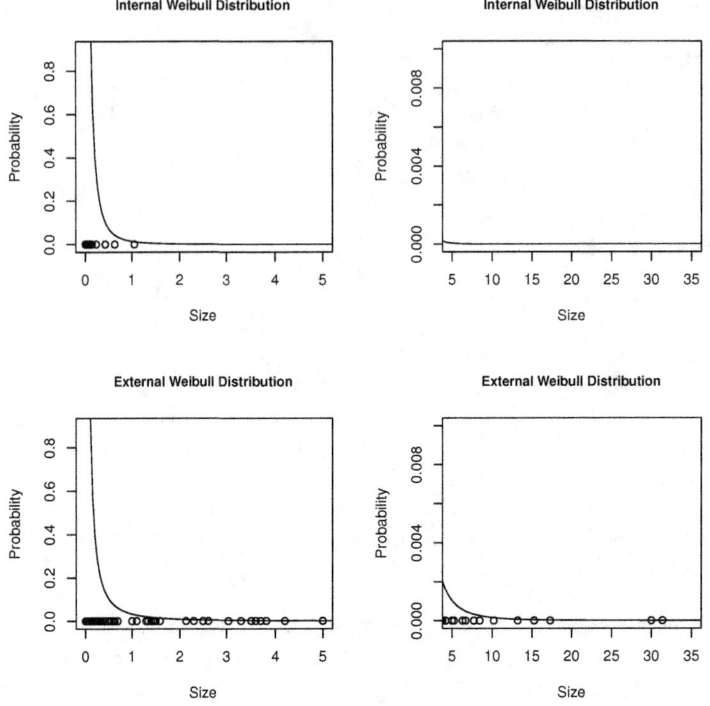

Figure 2.3 *The estimated density for internal losses (above) and external losses (below) operational risk data using the Weibull distribution.*

where c_* is a normalizing constant, and α, λ, and M are nonnegative parameters.

We have chosen this distribution because of its properties in the right-hand tail, that is, the area where large losses occur. When parameter λ equals 1 and the normalizing constant c_* equals $\frac{1}{2}\alpha$, the density of the original distribution is simply called the Champernowne pdf:

$$f(x) = \frac{\alpha M^\alpha x^{\alpha-1}}{(x^\alpha + M^\alpha)^2}$$

with cdf

$$F(x) = \frac{x^\alpha}{x^\alpha + M^\alpha}, \tag{2.2}$$

where α and M are positive parameters.

A characteristic feature of the Champernowne distribution is a convergence to a Pareto distribution in the tail, while looking more like a lognormal

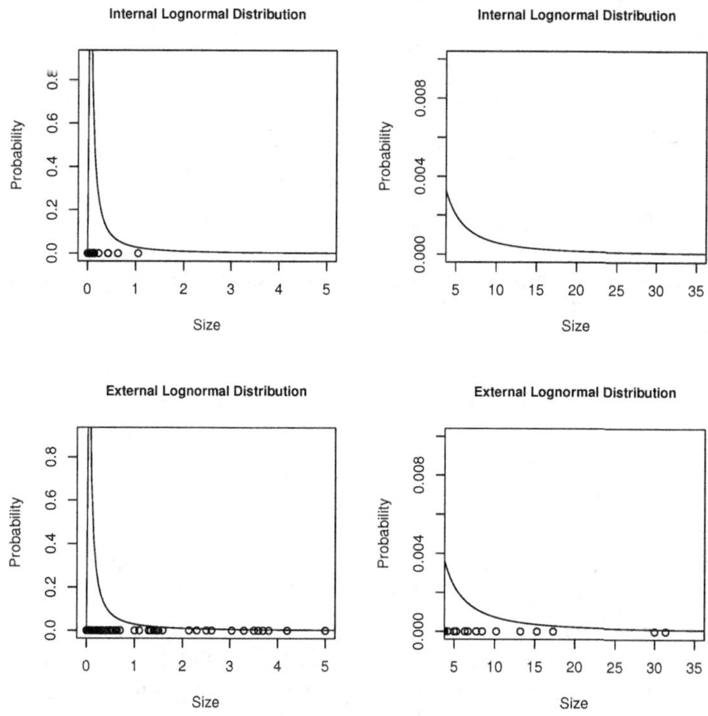

Figure 2.4 *The estimated density for internal losses (above) and external losses (below) operational risk data using the lognormal distribution.*

distribution near 0 when $\alpha > 1$. Its density is either 0 or infinity at 0 (unless $\alpha = 1$).

We did not choose to work with classical extensions of the Pareto distribution such as the generalized Pareto distribution (GPD) because we want to avoid choosing the starting point from where the domain is defined.

Distributions for extremes such as the generalized Pareto or the g-and-h distribution are suitable for large losses, but they are not recommended for the whole domain. Besides, they implicitly assume that a threshold has been chosen and that only those losses above the threshold need to be modeled. References can be found in the section on bibliographic notes at the end of this chapter.

In recent years there have been substantial improvements on the Champernowne distribution that includes parameter c. This new parameter ensures the possibility of a positive finite value of the density at 0 for all α.

The cdf of the generalized Champernowne[1] is defined as follows:

$$F_{\alpha,M,c}(x) = \frac{(x+c)^\alpha - c^\alpha}{(x+c)^\alpha + (M+c)^\alpha - 2c^\alpha} \qquad x \geq 0 \tag{2.3}$$

with parameters $\alpha > 0$, $M > 0$, and $c \geq 0$. Its density is

$$f_{\alpha,M,c}(x) = \frac{\alpha(x+c)^{\alpha-1}((M+c)^\alpha - c^\alpha)}{((x+c)^\alpha + (M+c)^\alpha - 2c^\alpha)^2} \qquad x \geq 0.$$

The generalized Champernowne distribution converges to a Pareto distribution in the tail. This means that if we define

$$g_{\alpha,M,c}(x) = \frac{\alpha\left(((M+c)^\alpha - c^\alpha)^{\frac{1}{\alpha}}\right)^\alpha}{x^{\alpha+1}},$$

then

$$\lim_{x\to\infty} \frac{f_{\alpha,M,c}(x)}{g_{\alpha,M,c}(x)} = 1.$$

The effect of the additional parameter c is different for $\alpha > 1$ and for $\alpha < 1$. The parameter c has some *scale parameter properties*: when $\alpha < 1$, the derivative of the cdf becomes larger for increasing c, and conversely, when $\alpha > 1$, the derivative of the cdf becomes smaller for increasing c. When $\alpha \neq 1$, the choice of c affects the density in three ways. First, c changes the density in the tail. When $\alpha < 1$, a positive c results in lighter tails, and the opposite when $\alpha > 1$. Second, c changes the density at 0, and a positive c provides a positive finite density at 0:

$$0 < f_{\alpha,M,c}(0) = \frac{\alpha c^{\alpha-1}}{(M+c)^\alpha - c^\alpha} < \infty \qquad \text{when } c > 0.$$

Third, c moves the mode. When $\alpha > 1$, the density has a mode, and positive c's shift the mode to the left. We therefore see that the parameter c also has a shift parameter effect. When $\alpha = 1$, the choice of c has no effect.

The role of parameter c is related to the mode of the density. Three examples ate presented in Figure 2.5. In the first row, the modified Champernowne distribution for fixed $\alpha < 1$ and $M = 2$ is presented. On the left, we have shown the cdf, and on the right, the corresponding pdf is presented. The cdf in the interval $[0, M)$ is larger for $c = 0$ than for $c > 0$, but in the interval $[M, \infty)$, the cdf is larger for $c > 0$ than for $c = 0$. In the corresponding density plot, $c > 0$ implies a lighter tail. In the second row, parameters are changed to $\alpha = 1$ and $M = 2$. In this case, parameter c does not influence the shape. In the last row, parameter $\alpha > 1$ and $M = 2$. In this case, the effect of parameter c is contrary

[1]This distribution is presented in Buch-Larsen's masters thesis, Laboratory of Actuarial Mathematics, Department of Mathematics and Statistics, University of Copenhagen.

to the effect observed in the graphs of the first row, so that the cdf when $c > 0$ is larger in the interval $[0, M)$ than the cdf fo $c = 0$. However, in the interval $[M, \infty)$, it is the other way round. In this case, a positive parameter c shifts the mode to the left.

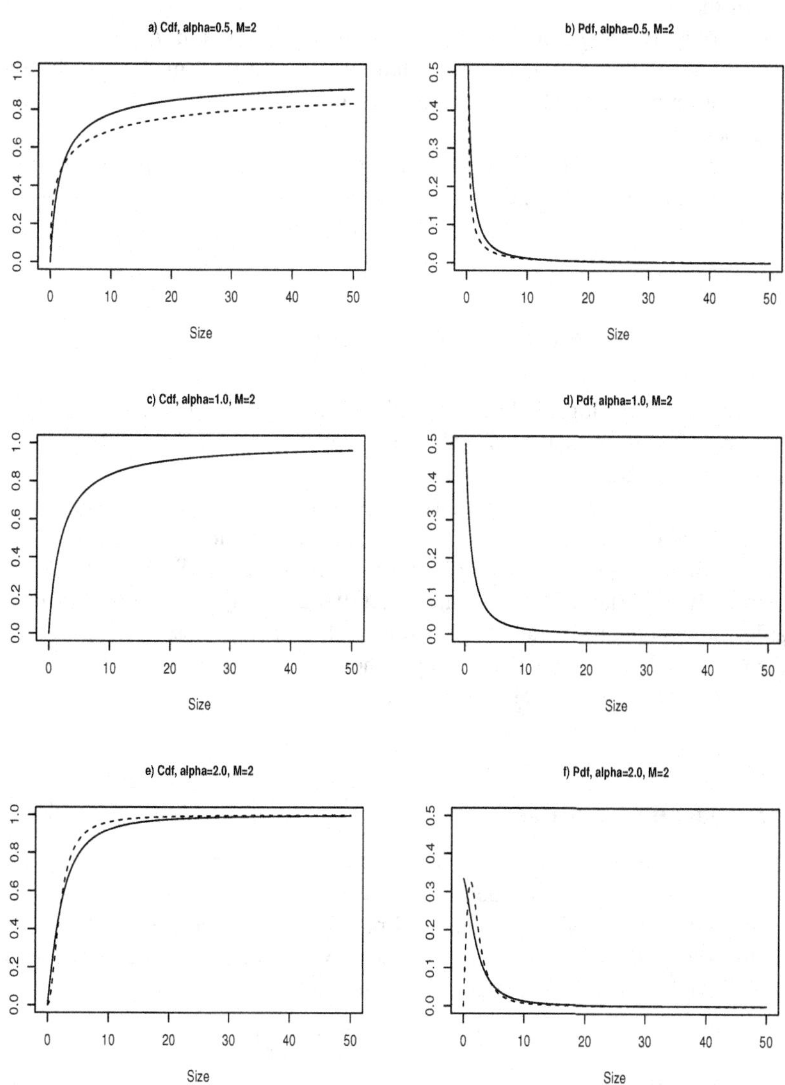

Figure 2.5 *Different shapes of the generalized Champernowne distribution. Positive c in solid line, and $c = 0$ in dashed line.*

A practical and simple to way to fit the generalized Champernowne distribution is to approximate the parameter step by step. We recommend M be estimated first and then move on to the other parameters, assuming that M, which corresponds to the median, is fixed and known. The reason for doing this is that the numerical procedures that lead to maximizing the likelihood do not always converge.

In the Champernowne distribution, we notice that $F_{\alpha,M,0}(M) = 0.5$. The same holds for the generalized Champernowne distribution: $F_{\alpha,M,c}(M) = 0.5$. This suggests that M can be estimated as the empirical median of the data set. The empirical median is a robust estimator, especially for heavy-tailed distributions.

Once parameter M is obtained, parameters (α, c) can be estimated by maximum likelihood, where the log-likelihood function is

$$\begin{aligned} l(\alpha,c) &= n\log\alpha + n\log((M+c)^\alpha - c^\alpha) + (\alpha-1)\sum_{i=1}^{n}\log(X_i+c) \\ &\quad -2\sum_{i=1}^{n}\log((X_i+c)^\alpha + (M+c)^\alpha - 2c^\alpha). \end{aligned} \tag{2.4}$$

For a fixed M, the log-likelihood function is concave and has a maximum.

The density estimated on both samples using maximum likelihood is presented in Figure 2.6.

Compared to the classical densities presented in the previous section, the generalized Champernowne distribution presents heavier tails that differ from zero in the domain areas where losses are still observed. We find this feature extremely important. Therefore the generalized Champernowne distribution is able to provide an overall fit both for small and large losses. It also avoids the use of extreme value distributions where the density is fitted only for large value and a threshold level must be chosen.

2.5 Quantile Estimation

Quantile estimation for the parametric distributions presented in the previous sections can be performed using the inverse of the distribution function evaluated at the maximum likelihood parameter estimates. So, if $F(x)$ is the distribution function of a random variable X and it is continuous and invertible, then the quantile at a level α is defined by

$$VaR_\alpha(X) = F^{-1}(\alpha),\ \alpha \in (0,1).$$

This section shows the quantile values from the distributions that have been presented in the previous sections. Figure 2.7 presents the values for each distribution estimated using the internal sample, while Figure 2.8 shows the quantile estimates based on the external sample.

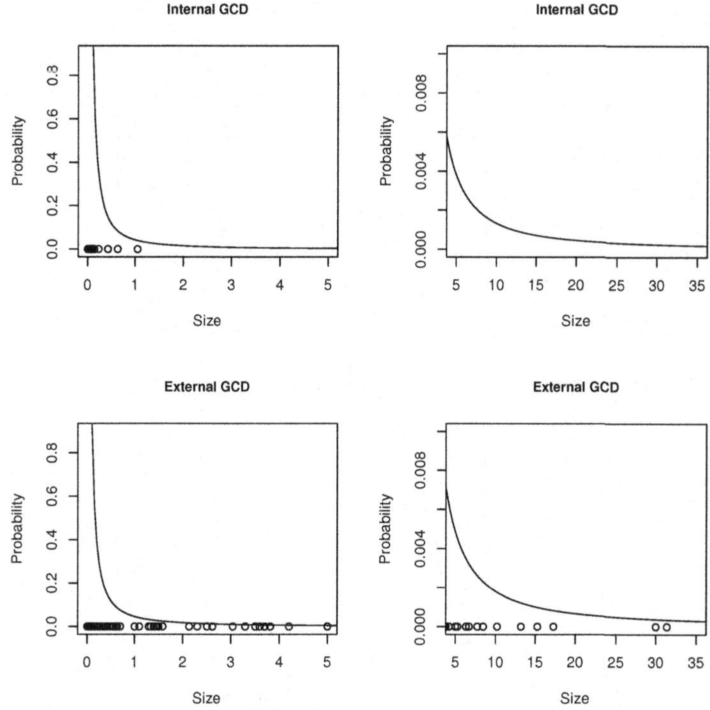

Figure 2.6 *The estimated density for internal losses (above) and external losses (below) operational risk data using the generalized Champernowne distribution (GCD).*

The results in Figures 2.7 and 2.8 should be interpreted with care since plots do not necessarily have the same y-axis scale. The main conclusion that can easily be derived from the plots is that the heavier the tail of the distribution, the higher the quantile value for a given probability α. So, the distributions that provide the largest quantiles is the generalized Champernowne distribution.

2.6 Further Reading and Bibliographic Notes

There are many bibliographic sources for continuous univariate distributions. One of the most well known is by Johnson et al. [52] and [53].

In [9] we can first attempt to use the generalized Champernowne distribution to the analysis of financial losses or insurance claim amounts.

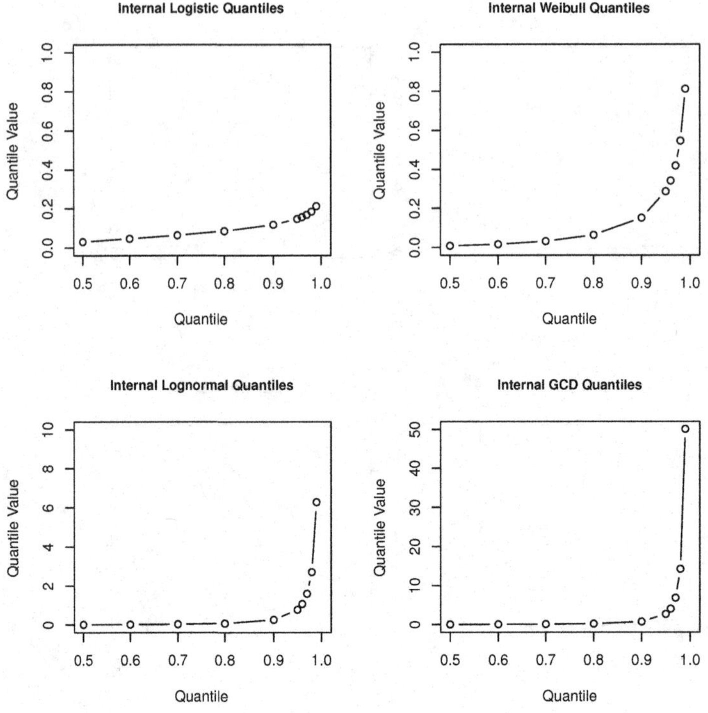

Figure 2.7: *Estimated quantile values from some distributions on the internal sample.*

However, this distribution was introduced there in the context of a semiparametric approximation that is further developed in Chapter 3 and which became well known after the contribution by [10].

The book by Klugman et al. [57] devotes a large part to univariate distributions for claims severities. References for extreme value distributions in this area include [31], [58], [64], [21], and [65].

There are many other references that, like [57], also include the analysis of claim counts. In [67], joint analysis of the total amount and the number of claims is modeled by conditionals.

A recent distribution that has promising potential value in the analysis of claim severities is the one described in [44], which was earlier developed by [68].

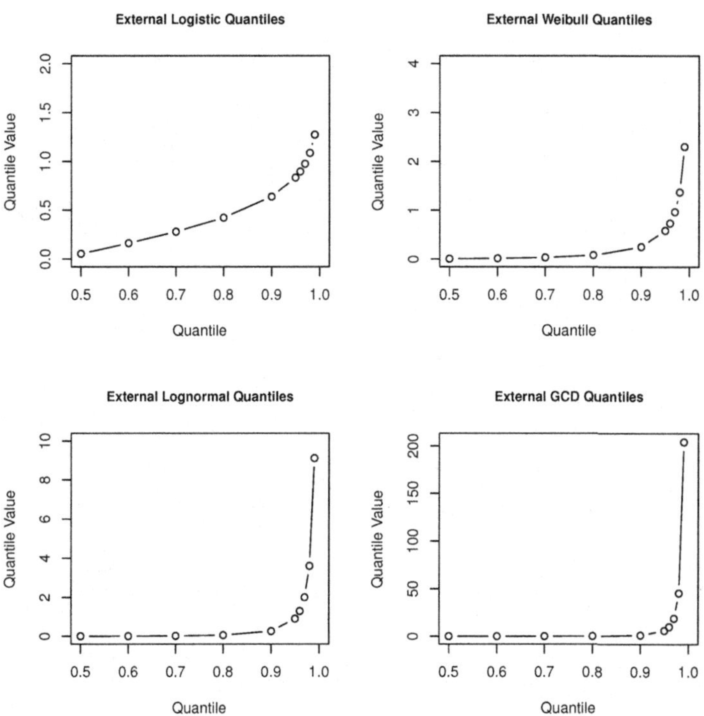

Figure 2.8 *Estimated quantile values from some distributions suggested on the external sample.*

Chapter 3

Semiparametric Models for Operational Risk Severities

3.1 Introduction

Traditional methods for studying severities or loss distributions use parametric models. Two of the most popular shapes are based on the lognormal and Pareto distribution. In the previous chapter we have shown how a parametric fit can be obtained, but this is not the only possible approach. In fact, the main reason why the parametric approach can be controversial is that a certain distribution is imposed on the data.

Alternatively, nonparametric smoothing is suitable for modeling operational loss data because it allows for more flexible forms for the density of the random variable that generates the information. The versatility of a nonparametric approach is convenient in our context because the shape of the density can be very different for small and large losses. In this chapter we present an approach based on a slightly adjusted version of the semiparametric transformed kernel density estimation method. We show that the estimation is easy to implement (including the often complicated question of choosing the amount of smoothing) and that the new method provides good and smooth estimates of the density in the upper tail. Parametric models have often been justified due to their simplicity and their ability to solve the problem of lack of smoothness of the empirical approach, so here we provide the practitioner with a nonparametric alternative to traditional parametric estimation techniques.

Operational risk analysts are interested in having good estimates of all the values in the domain range: small losses because they are very frequent, medium losses causing a dramatic increase of expenses (demanding liquidity), and large losses that may imply bankruptcy. We will study all severity sizes at the same time, and we will not break the domain into separate intervals. Actually, defining subintervals of the domain and fitting a distribution in every subinterval is an obvious approach when a unique parametric shape cannot be fitted across the whole domain range.

In finance and non-life-insurance, estimation of severity distributions is a fundamental part of the business. In most situations, losses are small, and extreme losses are rarely observed, but the size of extreme losses can have a substantial influence on the profit of the company. Standard statistical methodology, such as integrated error and likelihood, does not weigh small and big losses differently in the evaluation of an estimator. Therefore, these evaluation methods do not emphasize an important part of the error: the error in the tail.

As we have mentioned, practitioners often decide to analyze large and small losses separately because no single, classical parametric model fits all claim sizes. This approach leaves some important challenges: choosing the appropriate parametric model, identifying the best way of estimating the parameters, and determining the threshold level between large and small losses.

In this chapter we present a systematic approach to the estimation of loss distributions that is suitable for heavy-tailed situations.

Our approach is based on the estimation of the density of the severity random variable using a kernel estimation method. In order to capture the heavy-tailed behavior of the distribution, we propose to transform the data first and then to go back to the original scale using the reverse transform. This process is called transformation kernel density estimation. For the transformation, we use the generalized Champernowne cdf, which has been presented in the previous chapter. The combination of a nonparametric estimation with a previous parametric transformation of the data is the reason why we use the term semiparametric.

3.2 Classical Kernel Density Estimation

Classical kernel density estimation is a nonparametric method to approximate the probability density function of a random variable.

In order to give an understanding of what this method is about, we need to stress that we are estimating a density, that is, a function. So, if the random variable that we are studying is continuous and defined in the positive real line, then we need to provide an estimate of the density at every point in the variable domain. This is the reason why density estimation is a difficult problem. We do not estimate a single parameter but a whole set of unknown function values. An additional feature is that the density estimate must be nonnegative at every point, and it must integrate to one in order to preserve the basic properties of any probability density function.

Classical kernel density estimation is a method that will provide an estimator of the density function value at every point of the domain of the random variable. So, an expression will be given (depending of the sample observations and other constants) that can be evaluated at every point.

The basic idea is as follows. Our estimator will use the sample information that is in the neighborhood of the point where the function is estimated and ignore (or minimize the influence of) the information provided by the sample observations that are far away from that point. The way sample information is combined is determined using both the kernel function and the bandwidth or smoothing parameter.

For a random sample of n independent and identically distributed observations $X_1,...,X_n$ of a random variable X defined on a domain in R, the kernel

Table 3.1: *Expressions of most common kernel functions*

Kernel	$K(x)$
Uniform	$1/2$, if $\lvert x\rvert \leq 1$
Triangular	$(1-\lvert x\rvert)$, if $\lvert x\rvert \leq 1$
Epanechnikov	$(3/4)(1-x^2)$, if $\lvert x\rvert \leq 1$
Gaussian	$\left(1/\sqrt{2\pi}\right)e^{-(1/2)x^2}$, $\forall -\infty \leq x \leq +\infty$

density estimator of the true probability density $f(x)$ is

$$\hat{f}(x) = \frac{1}{nb}\sum_{i=1}^{n} K\left(\frac{x-X_i}{b}\right) \tag{3.1}$$

where $x \in \mathbb{R}$, b is the bandwidth constant parameter, and $K(\cdot)$ is the kernel function. Both b and $K(\cdot)$ have to be chosen. The bandwidth value b is used to control the amount of smoothing in the estimation so that the greater b, the smoother the estimated density curve. Section 3.4 in this chapter is devoted to the choice of the bandwidth in practical situations. The kernel function is usually a symmetric density with zero mean. Several kernel functions are available in the literature, and the most popular once are Gaussian, Epanechnikov, or triangular.

Table 3.1 shows the most popular kernel functions that are used in practice. Kernel functions are usually symmetric around zero. Note that the kernel function does not necessarily have to be a density itself, but if it is not a density, then the kernel estimate does not integrate to one. If the main aim is to estimate the density in one particular point $f(x)$, then it does not matter whether or not the kernel is a density, but when the aim is to estimate the density function over the whole domain, then one should use a kernel that is positive and integrates to one. The shape of the kernel does not significantly influence the final shape of the estimated density because it just determines the local behavior.

A common mistake for those unfamiliar with kernel density estimation is to think that using a certain kernel function implies that the final density estimate has the same shape as that of the kernel being used. This is not the case. For instance, a Gaussian kernel does not imply that the global density estimate is bell shaped.

The kernel function is just governing the way weights of sample observations are handled in the estimator. For instance, a Gaussian kernel will weight data according to a normal probability distribution, so even very distant observations from the point x where the density is estimated will have some minor influence on the estimation of the density at this given point x. A kernel defined on a bounded domain (like Epanechnikov's) will not use sample observations that are beyond its own domain.

When one is interested in estimating multivariate densities, the same procedure can be implemented. In the multivariate case, a simple generalization of (3.1) is done by means of the product kernels. More specifically, let us consider a random sample of n independent and identically distributed bivariate data (X_{1i}, X_{2i}), $i = 1, ..., n$, of the random vector $\mathbf{X} = (X_1, X_2)'$. Then the product kernel estimator of the bivariate density function can be expressed as

$$\hat{f}(x_1, x_2) = \frac{1}{nb_1b_2} \sum_{i=1}^{n} \mathbf{K}\left(\frac{x_1 - X_{1i}}{b_1}, \frac{x_2 - X_{2i}}{b_2}\right), \tag{3.2}$$

where b_1 and b_2 are bandwidths that, like in the univariate situation, are used to control the degree of smoothing. The function $\mathbf{K}\left(\frac{x_1-X_{1i}}{b_1}, \frac{x_2-X_{2i}}{b_2}\right)$ $= K\left(\frac{x_1-X_{1i}}{b_1}\right) K\left(\frac{x_2-X_{2i}}{b_2}\right)$ is the product kernel. This is the most simple way of considering multiple dimension kernel density estimation.

We will not use the multivariate approach in this chapter, but the method can easily be generalized from one dimension to two if this is of interest. We will continue working with the univariate case because the operational loss data that we focus on are assumed to have been generated by an one-dimensional random variable.

One important drawback of classical kernel density estimation is that it is biased in the mode when the density to be estimated is very heavy-tailed and the smoothing parameter is large. When there are some very large observations, the dispersion is large, and the bandwidth parameter is generally also quite big. Bias implies that the density is underestimated in the part of the domain with more mass. Alternatively, when a small bandwidth is used, the resulting density has a bumping shape at the tail due to the presence of some scarce large observations. This is the reason why we will introduce transformation kernel density estimation, a method that corrects for this bias. Since operational risk data are usually very heavy tailed, we do not recommend using classical kernel density estimation in this context. Instead, we will advocate the use of transformation kernel density estimation.

Figure 3.1 shows the estimation of the density of the internal and external operational risk data sets when the classical kernel density estimator is used. We know that, in this situation, the mode for small claims is underestimated, and we notice that, in the tail, there is some instability due to the local influence of very large losses. This is most significant for smaller-bandwidth-based estimates. For practical purposes the classical kernel density estimate is too irregular because usually practitioners like the smoother shape of the loss severity distribution.

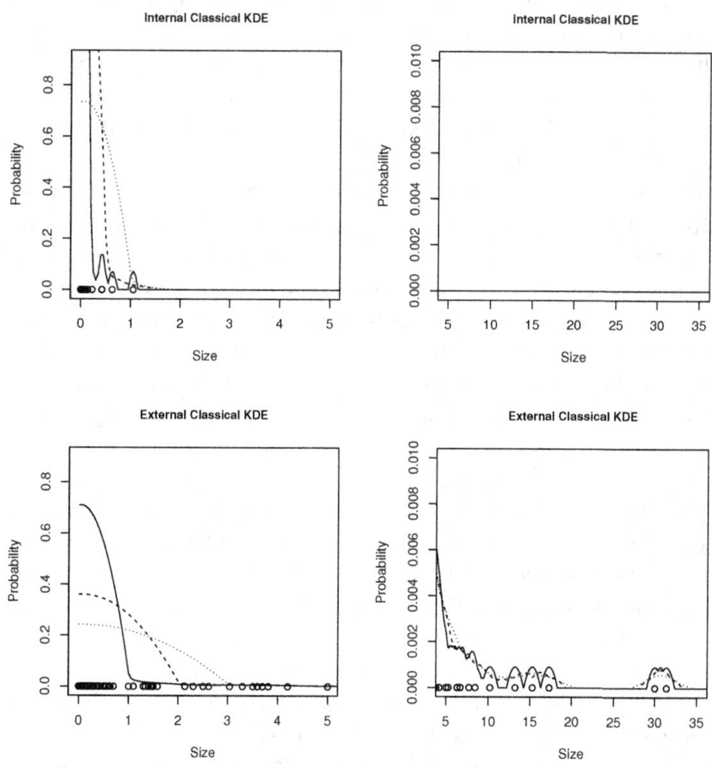

Figure 3.1 *Classical kernel density estimation (KDE) for the operational risk data sets. Above (below) is internal (external) operational risk data. Right-hand-side plots focus on large losses with adequate vertical scale units. Solid line corresponds to a bandwidth of size* $b = 0.1$, *dashed line corresponds to* $b = 0.5$, *and dotted line corresponds to* $b = 1$.

3.3 Transformation Method

We focus on our version of the semiparametric transformation approach to kernel smoothing, and we will see that the classical kernel estimation will substantially be improved. The method is called semiparametric because we use a parametric transformation function and then nonparametric kernel estimation. This chapter introduces a transformation method with a parametric cdf, and we advocate the semiparametric transformation method because it behaves very well when it comes to estimating severity distributions. Loss distributions, that is, densities associated to the severity of operational risk events, have typically one mode for the low loss values and then a long heavy tail. Existing results based on simulation studies have shown that the method is able to estimate all three possible kinds of tails, namely, the Fréchet type, the Weibull type, and the

Gumbel type. This makes the semiparametric transformation approach method extremely useful for operational risk analysis.

We will see that a specific feature of the semiparametric transformation principle is that it is also able to estimate the risk of a heavy-tailed distribution beyond the data. The reason is that extrapolation beyond the maximum observed value is straightforward.

When it comes to the actual implementation of the method, we only consider transformations that give a symmetric distribution. First, we asked ourselves the question, "Why is it possible to estimate beyond the data in density estimation when this is impossible in related estimation areas such as regression and hazard estimation?" The answer to this question seems to be that the density integrates to one. So, even if we do not have information on the tail, the restrictions on the density estimator provide us with valuable information. Recognizing the logic of this, we decided to estimate the entire density at once using the transformation method but forcing the transformation to result in a uniform density in the $[0,1]$ interval.

We emphasize that our purpose in this chapter is to construct a good overall method of estimating the density of risk severities that can be used both for the entire domain of the loss data and for its tail. Therefore, we present a general method for practitioners, applicable in a wide variety of situations, at least for a preliminary analysis of data.

In the following section, we study the properties of transformation kernel density estimation. Then we describe the use of the transformation. We apply these techniques to the operational loss data that are described in the previous chapter. We present the results and show how to estimate operational risk severity density functions in practice.

Transformations in kernel density estimation work as follows. Consider a twice-continuously-differentiable transformation. It could be a parametric or nonparametric cumulative distribution function or some other suitable function. The proposed density estimator is based on the principle of the transformation kernel density estimation method.

Let us assume that a sample of n independent and identically distributed observations $X_1,\ldots,X_n$ is available. We also assume that a transformation function $T(\cdot)$ is selected, and then the data can be transformed so that $Y_i = T(X_i)$ for $i = 1,\ldots,n$. We denote the transformed sample by $Y_1,\ldots,Y_n$.

The first step consists of transforming the data set with a function and afterwards estimating the density of the transformed data set using the classical kernel density estimator (3.1)

$$\widehat{f}(y) = \frac{1}{nb}\sum_{i=1}^{n} K\left(\frac{y-Y_i}{b}\right),$$

where K is the kernel function, b is the bandwidth, and Y_i for $i = 1,\ldots,n$ is the transformed data set. The estimator of the original density is obtained by a back-transformation of $\widehat{f}(y)$.

The transformed kernel density estimation method can be written as

$$\widehat{f}(x) = \frac{T'(x)}{n}\sum_{i=1}^{n} K_b\left(T(x) - T(X_i)\right), \tag{3.3}$$

where, as we mentioned, we have assumed that the transformations are differentiable and we have obtained this expression by simply using a change of variable-type random variable transformation. The superindex $'$ indicates the first derivative of a function. $K_b(\cdot) = \frac{1}{b}K(\cdot/b)$, where K refers to the kernel function and b is the bandwidth parameter.

3.3.1 Asymptotic Theory for the Transformation Kernel Density Estimator

In this section we summarize the asymptotic theory of the transformation kernel density estimator in general. Theorem 1 derives its asymptotic bias and variance.

Theorem 1 *Let $X_1,...,X_n$ be a random sample of observations from independent identically distributed random variables with density $f(x)$. Let $\widehat{f}(x)$ be the* transformation kernel density estimator *of $f(x)$ given in (3.3).*
Then the bias *and the* variance *of $\widehat{f}(x)$ are given by*

$$\begin{aligned}\mathbb{E}\left[\widehat{f}(x)\right] - f(x) &= \frac{1}{2}a_{21}b^2\left(\left(\frac{f(x)}{T'(x)}\right)'\frac{1}{T'(x)}\right)' + o(b^2),\\ \mathbb{V}\left[\widehat{f}(x)\right] &= \frac{1}{nb}a_{02}T'(x)f(x) + o\left(\frac{1}{nb}\right)\end{aligned}$$

as $n \to \infty$, where $a_{kl} = \int u^k K(u)^l\, du$.

Proof 2 *Let the transformed sample be $Y_i = T(X_i)$, $i = 1,...,n$. Then the density of the transformed variable from which the sample has been drawn is*

$$g(y) = \frac{f(T^{-1}(y))}{T'(T^{-1}(y))},$$

and let $\widehat{g}(y)$ be the classical kernel density estimator of $g(y)$:

$$\widehat{g}(y) = \frac{1}{n}\sum_{i=1}^{n} K_b(y - Y_i).$$

The expectation and variance of the classical kernel density estimator are

$$\mathbb{E}\left[\widehat{g}(y)\right] = g(y) + \frac{1}{2}b^2 a_{21} g''(y) + o(b^2), \tag{3.4}$$

$$\mathbb{V}\left[\widehat{g}(y)\right] = \frac{1}{nb}a_{02}g(y) + o\left(\frac{1}{nb}\right). \tag{3.5}$$

The transformation kernel density estimator can be expressed by the standard kernel density estimator:

$$\widehat{f}(x) = T'(x)\widehat{g}(T(x))$$

implying

$$\begin{aligned}\mathbb{E}\left[\widehat{f}(x)\right] &= T'(x)\mathbb{E}\left[\widehat{g}(T(x))\right] \\ &= T'(x)\left(g(T(x)) + \frac{1}{2}b^2 a_{21}\frac{\partial^2 g(T(x))}{\partial(T(x))^2} + o(b^2)\right).\end{aligned}$$

Note that $'$ *denotes differentiation with respect to x,* $g(T(x)) = \frac{f(x)}{T'(x)}$, $\frac{\partial g(T(x))}{\partial T(x)} = \left(\frac{f(x)}{T'(x)}\right)'\frac{1}{T'(x)}$, *and* $\frac{\partial^2 g(T(x))}{\partial(T(x))^2} = \left(\left(\frac{f(x)}{T'(x)}\right)'\frac{1}{T'(x)}\right)'\frac{1}{T'(x)}$, *which are used to find the mean of the transformation kernel density estimator,*

$$\mathbb{E}\left[\widehat{f}(x)\right] = f(x) + \frac{1}{2}b^2 a_{21}\left(\left(\frac{f(x)}{T'(x)}\right)'\frac{1}{T'(x)}\right)' + o(b^2). \tag{3.6}$$

The variance is calculated in a similar way:

$$\begin{aligned}\mathbb{V}\left[\widehat{f}(x)\right] &= (T'(x))^2\mathbb{V}\left[\widehat{g}(T(x))\right] \\ &= (T'(x))^2\left(\frac{1}{nb}a_{02}g(T(x)) + o\left(\frac{1}{nb}\right)\right) \\ &= \frac{1}{nb}a_{02}T'(x)f(x) + o\left(\frac{1}{nb}\right).\end{aligned} \tag{3.7}$$

It is known that the classical kernel density estimator follows a normal distribution asymptotically. We write $\sim$ *instead of "is distributed as," and therefore*

$$\sqrt{nb}\left(\widehat{g}(y) - \mathbb{E}\left[\widehat{g}(y)\right]\right) \sim N\left(0, a_{02}g(y)\right).$$

Then, since $\widehat{f}(x) = T'(x)\widehat{g}(y)$ *with* $y = T(x)$*, we get*

$$\sqrt{nb}\left(\widehat{f}(x) - \mathbb{E}\left[\widehat{f}(x)\right]\right) \sim N\left(0, a_{02}T'(x)f(x)\right).$$

3.3.2 *Transformation Method with a Cumulative Distribution Function*

We call this method a *semiparametric estimation procedure* when a parametrized transformation family is used for $T(\cdot)$. We show how to use a transformation based on the generalized Champernowne cdf, because it produces good results in most studied situations and it is straightforward to implement. Let

$$\mathsf{T} = \{T_\theta |\ \theta \in \Theta\}$$

be a set of twice continuously differentiable transformations. Θ can correspond to a parametric family of transformation functions, or just a set of θ parameters, or it can correspond to a bigger nonparametric class of functions.

For a given θ, the transformed density can be written using (3.3) as

$$\widehat{f}(x,\theta) = \frac{T'_{\theta}(x)}{n} \sum_{i=1}^{n} K_b\left(T_{\theta}(x) - T_{\theta}(X_i)\right),$$

where we have assumed that the transformations are differentiable. As before, the superindex $'$ indicates the first derivative of a function, and a random sample of n independent and identically distributed observations $X_1, ..., X_n$ is available. Note again that $K_b(\cdot) = \frac{1}{b}K(\cdot/b)$, where K refers to the kernel function and b is the bandwidth parameter.

Now, let $\widehat{\theta}$ be some consistent estimator for θ. If we are using the generalized Champernowne distribution, its parameters need to be estimated. Parameter estimation for the transformation is discussed later in this chapter. With estimated parameters, the resulting transformed density can be stated as

$$\widehat{f}(x,\widehat{\theta}) = \frac{T'_{\widehat{\theta}}(x)}{n} \sum_{i=1}^{n} K_b\left(T_{\widehat{\theta}}(x) - T_{\widehat{\theta}}(X_i)\right).$$

The asymptotic distribution is

$$\sqrt{nb}\left(\widehat{f}(x,\widehat{\theta}) - \mathbb{E}\left[\widehat{f}(x,\widehat{\theta})\right]\right) \sim N\left(0, a_{02}T'_{\theta}(x)\,f(x)\right).$$

For a parametric transformation $T(x) = T_{\theta}(x)$, if we assume that $\hat{\theta}$ is a square-root-n consistent estimator of θ, then it follows that the asymptotic distribution of $\widehat{f}(x)$ with parametric estimated transformation $T_{\widehat{\theta}}(x)$ equals the asymptotic distribution of $\widehat{f}(x)$ with parametric transformation $T_{\theta}(x)$.

The semiparametric transformation method is approximately equivalent to the classical kernel density estimation method with a variable bandwidth. If a transformation is selected with $T'_{\theta}(x)$ going to zero for x going to infinity, the parametric transformation method therefore approximately ensures a bandwidth going to infinity in the tail, resulting in a very stable tail behavior of the estimator. Moreover, the difference between the variable bandwidth kernel density estimation procedure and the transformation method is that the latter uses a local smoothing of the estimator, ensuring that the final estimator integrates to one. The conclusion is that the semiparametric transformation method seems to be useful when estimating densities over the entire domain including the tail. When introducing the transformation family, we will see that, when this parametric transformation is applied to real operational risk data, it provides estimators with very stable tail behavior.

Since we want to explain the cooperation between bandwidth selection and transformation, we proceed now to discuss the choice of bandwidth. In combination with a transformation, the bandwidth encompasses a trade-off between the information in the data and the information in the transformation. In fact, our transformation can be based on

- External data. When the transformation is based on external data, then it represents the way we put external information into our model.
- Parametric model. If the transformation is based on a parametric model, it represents the way the parametric knowledge gets into the model.
- Internal data. A classical nonparametric approach to the cdf of internal data can be used as a smoothing trick and as a starting point for transformation.

3.4 Bandwidth Selection

The choice of bandwidth is an essential point when using nonparametric smoothing techniques. The bandwidth itself could be thought of as a scaling factor that determines the width of the kernel function, and thereby it controls how wide a probability mass is provided in the neighborhood of each data point.

In combination with the transformation, the bandwidth excerpts a very different effect in the original scale. For instance, a small bandwidth that is applied to a transformation that has compressed the original data quite considerably, would clearly cover a wide interval in the original situation (untransformed scale). Conversely, a large bandwidth in a transformation that has widened a subdomain in the original scale may be equivalent to a small bandwidth that is used in the original scale. So, a constant bandwidth applied after the transformation to obtain a kernel estimate has the same effect as implementing a variable bandwidth in the original scale with no transformation.

Different bandwidth selection methods for improving the performance of nonparametric kernels have been suggested. From the fact that a transformation method will be implemented, bandwidth selection has to be adapted to our modeling situation.

The simplest approach to bandwidth choice is the method known as Silverman's rule-of-thumb bandwidth, also termed normal scaled bandwidth. The idea for an optimal bandwidth is to minimize the Mean Integrated Squared Error (MISE). This is solved by inserting the asymptotic result for interior bias and variance of the classical kernel density estimation, ignoring the effect occurring from boundaries, approximated as

$$\frac{a_{02}}{nb} + \frac{1}{4} b^4 a_{21}^2 \int f''(x)^2 dx.$$

Consequently, if we differentiate this with respect to b, we obtained the theoretical optimal choice

$$b = \left(\frac{a_{02}}{a_{21}^2 n \int f''(x)^2 dx} \right)^{1/5}.$$

Here everything is known except for the density f. As noted before, $a_{kl} = \int u^k K(u)^l \, du$, and K is the kernel function.

A straightforward assumption is that the unknown f is the density of a normal random variable with scale parameter σ and that the Epanechnikov kernel is used, and then the bandwidth estimate is

$$b = \widehat{\sigma} \left(40\sqrt{\pi}\right)^{1/5} n^{-1/5},$$

where $\widehat{\sigma}$ is the standard deviation that is estimated from the sample. Note that the bandwidth is proportional to $n^{-1/5}$, meaning that increasing sample size brings decreasing optimal bandwidth.

Extensions to the above approach can improve bandwidth selection in specific situations.

3.5 Boundary Correction

In nonparametric smoothing techniques, boundary correction is needed when data are observed in a limited interval even if it is just on the left- or right-hand side and the density is strictly positive in the bounds. Losses are positive, so there is naturally a limit in the left of the domain. When using the transformation method, boundary correction is necessary because data are transformed into the $[0,1]$ interval using a cumulative distribution function.

This section describes a simple method to address the issues of boundary bias induced by transformation to bounded support. The approach for a one-method-fits-all estimator starts with transforming losses with the generalized Champernowne distribution into the interval $[0,1]$ such that it approximates a uniform distribution. Thereafter, application of nonparametric techniques allow an improved fit to the data. This procedure is termed semiparametric, and provides a link between the parametric and the nonparametric methodologies. The advantages of utilizing parametric transformation can be summarized as follows: when data are sparse, we do not want to rely entirely on nonparametric estimation, and therefore, the estimation is close to the parametric model. As sample size increases, the estimator converges to a nonparametric model. Further, it solves effectively the generally known weakness of kernel density estimation in the tail of a distribution. However, parametric transformation leaves us with challenges surrounding boundary bias. Since the classical kernel density estimation was primarily developed with unbounded support and thereby have no knowledge of the boundaries, probability mass occurs outside the support. The estimator then would not integrate to one.

The parametric transformation approach to kernel smoothing is based upon the transformation $T_\theta : R \rightarrow [0,1]$, achieved by

$$Y_i = T_{\widehat{\theta}}(X_i),$$

where X_i, $i = 1, \ldots, n$ are losses from some group of data, and $\widehat{\theta}$ are the estimated parameter vector belonging to the generalized Champernowne distribution.

The classical kernel density estimation with the underlying parametric transformation is then presented as

$$\widehat{g}(y) = \frac{1}{n}\sum_{i=1}^{n} K_b\left(y - Y_i\right), \tag{3.8}$$

where $y \in [0,1]$ and K denotes a probability density function symmetric about zero with support $[-1,1]$, and let $K_b(\cdot) = \frac{1}{b}K(\frac{\cdot}{b})$ and bandwidth $b > 0$.

We now define the following functions giving the asymptotic properties of the kernel density estimator around boundaries,

$$a_{kl}(y,b) = \int_{\max\{-1,(y-1)/b\}}^{\min\{1,y/b\}} u^k K(u)^l du, \text{ for } y \in [0,1]. \tag{3.9}$$

Note that $a_{01}(y,b) = 1$ and $a_{11}(y,b) = 0$ for the interior points y, defined in the interval $[b, 1-b]$. At the boundary points belonging to the intervals $[0,b)$ and $(1-b,1]$, $a_{01}(y,b)$ and $a_{11}(y,b)$ take nontrivial values.

To describe the asymptotic properties, we let $Y_1,\ldots,Y_n$ be *iid* random observations of the transformed variables with density g and suppose that g has two continuous derivatives everywhere, and $n \to \infty$, $b \to 0$, and that $nb \to \infty$ holds. Then the asymptotic bias

$$\mathbb{E}\left[\widehat{g}(y) - g(y)\right] =$$

$$\begin{cases} \frac{1}{2}b^2 a_{21}(y,b)g''(y) + O(b^2) & y \in [b, 1-b] \\ (a_{01}(y,b) - 1)\,g(y) - b a_{11}(y,b)g'(y) + O(b) & y \in [0,b) \text{ and } y \in (1-b,1], \end{cases}$$

and the asymptotic variance

$$\mathbb{V}\left[\widehat{g}(y)\right] = \begin{cases} \frac{a_{02}(y,b)}{nb} g(y) + O\left(\frac{1}{nb}\right) & \text{for } y \in [b, 1-b] \\ \frac{a_{02}(y,b)}{nba_{01}(y,b)^2} g(y) + O\left(\frac{1}{nb}\right) & \text{for } y \in [0,b) \text{ and } y \in (1-b,1]. \end{cases}$$

We note that, in the interior domain, we have bias of $O\left(b^2\right)$, but reaching the boundaries, the bias becomes $O\left(b\right)$. The variance is of $O((nb)^{-1})$ with a multiplier $a_{02}(y,b)$, defined through (3.9). Note that the variance depends on both sample size n and bandwidth value b, while bias depends only on b. Consequently, if b is chosen to be too small, we are punished by increasing variance for finite sample size. Also note that the variance increases when reaching the boundaries since $a_{01}(y,b) < 1$.

The final step in the transformation process is to backtransform to the original axis. This is done by transforming the smoothed distribution (3.8) of the transformed data with $T_{\widehat{\theta}}^{-1}(y)$. Then the semiparametric transformation process could be summarized in one explicit expression

$$\widehat{f_{o.s}}(x) = \frac{T'_{\widehat{\theta}}(x)}{n} \sum_{i=1}^{n} K_b\left(T_{\widehat{\theta}}(x) - T_{\widehat{\theta}}(X_i)\right),$$

where $x \in R_+$. Note that the indexation $o.s$ above stands for original scale.

We consider one possible alternative kernel density estimation to achieve consistency, in contrast to (3.8), at and near the boundaries. The first-presented kernel density estimator is the most simple one obtained by renormalizing the classical kernel density estimator (3.8). That is, for each boundary point, the classical kernel density estimator is divided by the area from the outside mass. This method is equivalent to the classical kernel density estimator for the interior points, meaning that the correction only affects the estimator for overspilling the boundaries and thereby provides a unit integration. The local constant kernel density estimator is defined as

$$\widehat{g_{lc}}(y) = \frac{1}{a_{01}(y,b)n} \sum_{i=1}^{n} K_b\left(y - Y_i\right) = \frac{\widehat{g}(y)}{a_{01}(y,b)}, \tag{3.10}$$

where $\widehat{g}$ is defined in (3.8). Similarly, we can backtransform into the original scale so that the transformation kernel density estimation with boundary correction is

$$\widehat{f_{o.s}}(x) = \frac{T'_{\widehat{\theta}}(x)}{a_{01}(T_{\widehat{\theta}}(x),b)n} \sum_{i=1}^{n} K_b\left(T_{\widehat{\theta}}(x) - T_{\widehat{\theta}}(X_i)\right).$$

3.6 Transformation with the Generalized Champernowne Distribution

In this section we will make a detailed derivation of the transformation density estimator based on the generalized Champernowne distribution. This distribution was presented in Chapter 2, where it was used as a parametric alternative to fit the density.

Initially, the original data observations are transformed using the cdf of the generalized Champernowne distribution. Then a classical kernel density estimator for the transformed data set is applied and, finally, the result is back-transformed to the original scale.

Let X_i, $i = 1,...,n$, be independent observations obtained from a stochastic variable X with an unknown cdf $F(x)$ and pdf $f(x)$. Let $T_{\alpha,M,c}(\cdot)$ be the generalized Champernowne distribution function used for the transformation, with parameters $\theta = (\alpha, M, c)$. So, the cdf and pdf are

$$T_{\alpha,M,c}(x) = \frac{(x+c)^{\alpha} - c^{\alpha}}{(x+c)^{\alpha} + (M+c)^{\alpha} - 2c^{\alpha}} \qquad x \geq 0,$$

$$T'_{\alpha,M,c}(x) = \frac{\alpha(x+c)^{\alpha-1}((M+c)^{\alpha} - c^{\alpha})}{((x+c)^{\alpha} + (M+c)^{\alpha} - 2c^{\alpha})^2} \qquad x \geq 0,$$

with parameters $\alpha > 0$, $M > 0$ and $c \geq 0$.

The following four steps of the estimation procedure describe in detail the transformation kernel density estimator of $f(x)$.

1. Calculate the parameters $\left(\widehat{\alpha}, \widehat{M}, \widehat{c}\right)$ of the generalized Champernowne distribution as described in the previous chapter to obtain the transformation function.
2. Transform the data set X_i, $i = 1,...,n$, with the transformation function, $T_{\widehat{\alpha},\widehat{M},\widehat{c}}(\cdot)$ that corresponds to the estimated cdf of the generalized Champernowne distribution:

   $$Y_i = T_{\widehat{\alpha},\widehat{M},\widehat{c}}(X_i), \qquad i = 1,...,n.$$

   The transformation function transforms data into the interval $[0,1]$, and the parameter estimation is designed to make the transformed data as close to a uniform distribution as possible.
3. Calculate the classical kernel density estimator on the transformed data, $Y_i, i = 1,...,n$, including a boundary correction:

   $$\widehat{f}_{trans}(y) = \frac{1}{n\, a_{kl}(y,b)} \sum_{i=1}^{n} K_b(y - Y_i),$$

   where $K_b(\cdot) = \frac{1}{b}K(\cdot/b)$, and $K(\cdot)$ is the kernel function. The classical expression arises when $a_{kl}(y,b) = 1$. Boundary correction

   $$a_{kl}(y,b) = \int_{\max(-1,-y/b)}^{\min(1,(1-y)/b)} u^k K(u)^l \, du$$

   is required because the Y_i are in the interval $[0,1]$ so that we need to divide by the integral of the part of the kernel function that lies in this interval. The easier boundary correction $a_{01}(y,b)$ is defined as

   $$a_{01}(y,b) = \int_{\max(-1,-y/b)}^{\min(1,(1-y)/b)} K(u)\, du.$$

   More details on how to perform boundary corrections will be found at the end of this chapter. A note on how to choose the bandwidth parameter b is described in Section 3.4.
4. The classical kernel density estimator of the transformed data set results in the transformation kernel density estimator based on the Champernowne

distribution on the transformed scale. Therefore, the estimator of the density of the original severities, X_i, $i = 1,...,n$ is

$$\widehat{f}(x) = \frac{\widehat{f_{trans}}\left(T_{\widehat{\alpha},\widehat{M},\widehat{c}}(x)\right)}{\left|\left(T^{-1}_{\widehat{\alpha},\widehat{M},\widehat{c}}\right)'\left(T_{\widehat{\alpha},\widehat{M},\widehat{c}}(x)\right)\right|}.$$

The expression of the semiparametric transformation density estimator (including the boundary correction introduced in the third step) is:

$$\widehat{f}(x) = \frac{T'_{\widehat{\alpha},\widehat{M},\widehat{c}}(x)}{n\, a_{01}(T_{\widehat{\alpha},\widehat{M},\widehat{c}}(x), b)} \sum_{i=1}^{n} K_b(T_{\widehat{\alpha},\widehat{M},\widehat{c}}(x) - T_{\widehat{\alpha},\widehat{M},\widehat{c}}(X_i)).$$

3.7 Results for the Operational Risk Data

In this section we will apply our semiparametric estimation method to two data sets. The first data set is internal operational risk data, and the second one is external operational risk data. We displayed in Figure 3.1 the result of classical kernel density estimation. In the top panel internal operational risk data are shown, and the external data are presented below. Right-hand-side plots focus on large losses with adequate vertical scale units. The picture shows that the estimated density in the tail is bumpy.

Estimation of the parameters in the generalized Champernowne distribution function are, for internal losses, $\widehat{\alpha}_1 = 0.5595$, $\widehat{M}_1 = 0.0136$, and $\widehat{c}_1 = 0.0000$, and for external losses, $\widehat{\alpha}_2 = 0.4699$, $\widehat{M}_2 = 0.01157$ and, $\widehat{c}_2 = 0.0000$, respectively. We notice that $\widehat{\alpha}_1 > \widehat{\alpha}_2$, which indicates that the data set for external losses has a heavier tail than the data set for internal losses when fitting the parametric density.

Figure 3.2 shows the histograms for the transformed data on operational risk on the left for internal data and on the right for external data. The estimated generalized Champernowne distribution is being used for transforming the original samples.

Figure 3.3 presents the local constant boundary correction classical kernel density estimator of the transformed data separated by the two internal and external sources. The estimated bandwidths that are used are $b_1 = 0.2637$ and $b_2 = 0.1669$, respectively.

When plotting the estimators on the original scale, we see the importance of considering the semiparametric method based on the generalized Champernowne distribution. The final results in the original scale are shown in

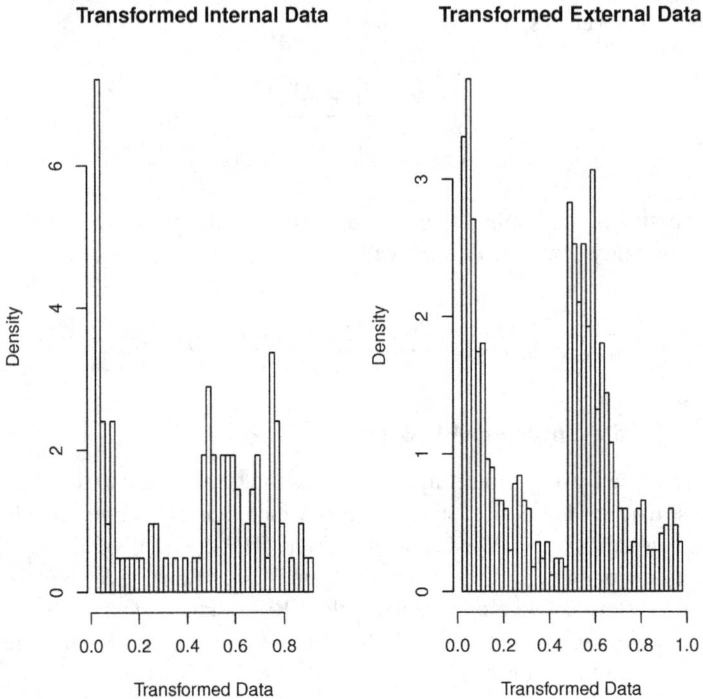

Figure 3.2 *Transformed internal and external operational risk data with the estimated cdf of the generalized Champernowne distribution.*

Figure 3.4. In order to show the shape of the estimated density in the tail, we show the part of the plot corresponding to large losses on the right-hand-side.

3.8 Further Reading and Bibliographic Notes

A central reference for this chapter is Buch-Larsen [9], which was inspired by the research previously done by Bolancé et al. [4]. The former derived from the joint work with the later [4] and lead to the publication of Buch-Larsen et al. [10]. We acknowledge that these three references and some working papers that were produced in that period of time form the core of the research developed in the present book.

The standard actuarial curriculum considers kernel density estimation as a fundamental topic (see [57]). Insurance practitioners and actuaries, however, have not made extensive use of nonparametric methods in this field. In the context of graduating mortality, we should note the papers by Gavin et al. ([39], [40]). Our opinion is that actuaries will only use a method that is particu-

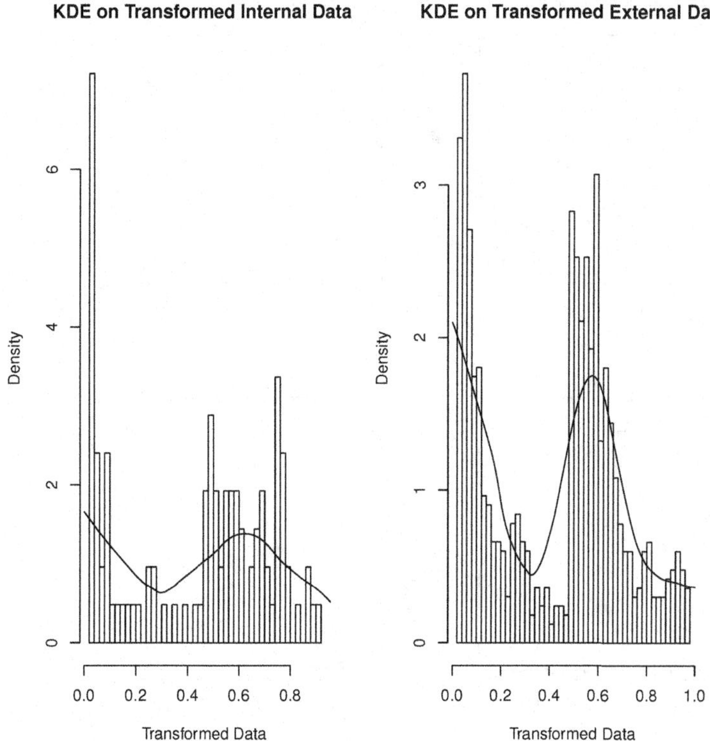

Figure 3.3 *Transformed internal (left) and external (right) operational risk data with the generalized Champernowne distribution and the respective kernel density estimation (KDE) with local constant boundary correction.*

larly good at estimating the density in the tails, a question that classical kernel smoothing fails to adequately address.

Transformation kernel density estimation was proposed by Wand et al. [74]. The semiparametric transformation approach to kernel smoothing introduced is an alternative approach to the standard kernel estimator and the multiplicative bias correction method presented by Hjort and Glad [50] and Jones et al. [54].

A simulation study and a different application of the method presented in this chapter can be found in [10]. The simulation study shows that the semiparametric transformation approach is able to estimate all three possible kinds of tails, as defined in [31] (p. 152). This makes this method extremely useful for operational risk analysis. The semiparametric estimator with shifted power transformation introduced by Wand et al. [74] improves the classical

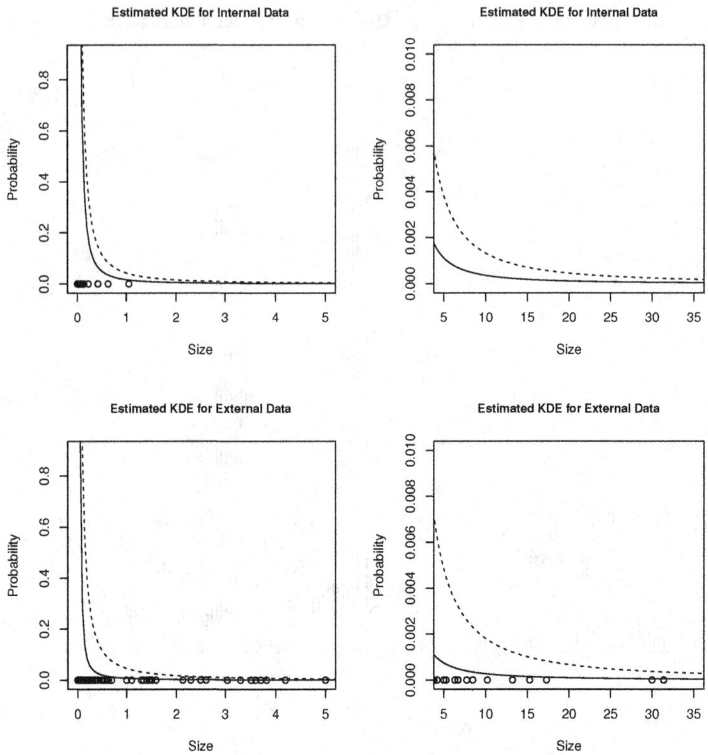

Figure 3.4 *Internal (above) and external (below) operational risk data in the original scale. Solid line represents the semiparametric transformed kernel density estimation (KDE) with the generalized Champernowne distribution and a local constant boundary correction. The dashed line corresponds to a pure parametric fit of a generalized Champernowne distribution. Right-hand-side plots focus on large losses with adequate vertical scale units.*

kernel density estimator by applying a transformation and suggested the shifted power transformation family. Bolancé et al. [4] improved the parameter selection of the shifted power transformation for highly skewed data by proposing an alternative algorithm. The semiparametric estimator with the Johnson family transformation function was studied by [77]. Hjort and Glad [50] advocated a semiparametric estimator with a parametric start, which is closely related to the bias reduction method described by [54]. The Möbius-like transformation was introduced by [20]. In contrast to the shifted power transformation, which transforms $(0,\infty)$ into $(-\infty,\infty)$, the Möbius-like transformation transforms $(0,\infty)$ into $(-1,1)$, and the parameter estimation method is designed to avoid boundary problems.

The parameter set, Θ, in the transformation function can correspond to a parametric family of transformation functions or just a set of parameters as in [74] and also in [77], or it can correspond to a bigger nonparametric class of functions, as in [66] and [51].

More details on the semiparametric transformation kernel density estimation can be found in [4], where they also show that the parametric transformation method is approximately equivalent to the semiparametric multiplicative kernel density estimation method of [50] with a variable bandwidth equal to a constant time $\left(T_{\theta}'(x)\right)^{-1}$. Extensions to the multivariate case can be found in [8]. A work that relates the multivariate case with risk measurement is [13].

Fan and Gijbels [34] discussed the importance of estimating a correct bandwidth and explored local and global variable bandwidths. They introduced a function into the kernel estimator, thereby letting the amount of smoothing depend on the local exposure. They recommended this approach for improved performance when exposures vary with high curvature, which is not the case when data are transformed to approximately uniform. Another appealing bandwidth method is the least-squares-cross-validation. The basic idea is to minimize the Integrated Squared Error (ISE) for the leave-one-observation-out estimator, which could be estimated with any of the kernel density estimators. The minimizing process is obtained by utilizing the grid search method, which seeks minimum for an estimation of ISE repeatedly for different bandwidth choices. This method is discussed in more detail in [70] and [73]. Wand and Jones [73] also provide an extended review of density estimation at the boundaries. Gustafsson and Nielsen [47] provide more insights into the transformation method and boundary correction.

The proof of asymptotic results for the bandwidth selection can be found in [70]. Bolancé et al. ([5] and [6]) also addressed other transformations and bandwidth selections, For instance, a double transformation can be used in order to apply automatic bandwidth selection that is optimal.

Finally, [76] provides formal proofs of the asymptotic behavior of the transformation kernel estimator when consistent square-root-n parameter estimates are used for the transformation function.

Chapter 4

Combining Operational Risk Data Sources

4.1 Why Mixing?

When measuring operational risk, having enough data is one of the main concerns. Collecting data for operational risk is difficult. Managers might be worried first about whether the reference period is homogeneous with the current period, and second, whether all operational failures have been recorded. However, there is a third issue that usually escapes from the control of operational risk managers, and that is the amount of events that has been observed. If there are too many, then the manager might focus on frequency and tend to be more concerned about reducing operational risk events, but if there are too few events, then a statistical problems arises because the sample size of the available data may not be sufficient to draw conclusions that are statistically sound. Some of these issues will be dealt with in the chapter that discusses underreporting; however, we now want to concentrate on the combination of data.

Operational losses occur randomly and affect one company and not another one, but it could have been otherwise. Therefore, occasionally, the regulator may suggest that internal data should be mixed with external data.

Calculating loss distributions for operational risk by only using internal data often fails to capture potential risks and unprecedented large loss amounts that have huge impact on the capital. Conversely, calculating loss distribution by only using external data provides capital estimation that is not sensitive to internal losses. For instance, an estimate that is exclusively based on external information would not increase despite the occurrence of a large internal loss and would not decrease despite the improvement of internal controls.

In general, external data should be combined with internal data on operational risk in order for a company to

1. obtain extra and valuable information of events that could have occurred also
2. increase sample size and guarantee a solid statistical basis for prediction of future events
3. measure the distance between internal and external behavior.

All three objectives would improve risk management but would also raise questions such as the fact that a mixture of internal and external data is not only describing what has been happening in the company but what has been happening in other companies as well. The population that is studied then changes. Another question is how the mixture should be done, how much importance should be given to each of the data sets, and whether there is a unique way of mixing internal and external sources. The answer is that there is not a single way of pooling together the information, and results may vary depending on the statistical hypothesis.

When mixing internal data with another source, we will assume that there may be restrictions with internal data so it makes sense to mix them with an external data set, or that the external data set may be set up in terms of a scenario and not as a data set.

In this chapter we discuss a method to estimate operational risk loss distribution that considers both internal and external data. The procedure is direct and is useful in obtaining a compromise between the two sources.

4.2 Combining Data Sources with the Transformation Method

The model will incorporate information both from internal and external data. We first go through an intuitive and simplified example on how *prior* knowledge can be incorporated into modern estimation techniques.

Let us first assume that our data and densities are defined on the $[0,1]$ interval. This makes our intuition easier because the domain is bounded.

In this oversimplified example, we consider a stochastic variable with density h on interval $[0,1]$, and let h_θ be another density on $[0,1]$ that is defined by *prior* knowledge. We will focus on estimating h and we consider *prior* knowledge as given by h_θ, which is assumed to be known.

Let us define $g_\theta = h/h_\theta$, which should be close to the uniform, and let us estimate g_θ by an estimator $\hat{g}_\theta$. Then the estimator $\hat{h}$ of h based on *prior* knowledge is $\hat{h} = h_\theta \cdot \hat{g}_\theta$.

To estimate g_θ we take advantage of nonparametric smoothing techniques, where we consider the simplest smoothing technique available, which is classical kernel smoothing discussed in the previous chapter. Note that, in this way, in order to obtain an estimation of h, we have changed the problem from estimating the completely unknown and complicated density h to the simpler g_θ that is close to the density of a uniform distribution when h_θ represents accurate *prior* knowledge.

Now let $\hat{g}_\theta$ be a kernel density estimation function estimating g_θ. The nature of kernel smoothing is such that it smooths a lot when data is sparse. In our case this means that $\hat{h}$ will be close to h_θ for very small sample sizes. When data are more abundant, less kernel smoothing will take place, and the complicated features of g_θ can be visualized. Therefore, kernel smoothing incorporating *prior* knowledge has many similarities to Bayesian methodology or credibility theory that also use global information as *prior* knowledge when data are scarce, and a local estimation when data are abundant.

Section 4.3 will explain how to implement the ideas for unbounded non negative domain data. We will then use the transformation approach explained in Chapter 3 to change the variables to be defined on the $[0,1]$ interval, and we will also introduce boundary correction.

Our purpose is to use *prior* knowledge from external data to improve our estimation of the internal data distribution. The underlying concept of the new mixing model transformation approach reflects the simple intuition in the illustration above on how to include *prior* knowledge on $[0,1]$ densities. In our case we wish to estimate loss distributions on the entire positive real axis. This is done by estimating a distribution on the external data, and we will use the generalized Champernowne distribution for this purpose. The generalized Champernowne distribution estimated from the external data plays the same type of role as the prior knowledge h_θ above. The external data represent our *prior* knowledge.

4.3 The Mixing Transformation Technique

For simplicity, we introduce superindex I to indicate an internal data source, and we use superindex E to denote an external source. The transformed data set will have a density equal to the true internal density divided by the estimated external density distribution transformed such that the support is on $[0,1]$.

When the external data distribution is a close approximation to the internal data distribution, we have simplified the problem to estimating something close to a uniform distribution. When this estimation problem has been solved through kernel smoothing, we backtransform to the original axis and find our final mixing estimator.

Let $X_1^I, X_2^I, \ldots, X_{n_I}^I$ be a sequence of n_I collected internal losses, and let $X_1^E, X_2^E, \ldots, X_{n_E}^E$ be a sequence of n_E external reported losses. Then $X_1^E, X_2^E, \ldots, X_{n_E}^E$ represent *prior* knowledge that should enrich a limited internal data set. We assume that scales are comparable and a filtration has been made, so that the external information has already been prepared to be directly mixed with the internal data.

The internal data sample $X_1^I, X_2^I, \ldots, X_{n_I}^I$ is then transformed with the estimated external cumulative distribution function, that is, using $T_{\widehat{\alpha}^E, \widehat{M}^E, \widehat{c}^E}(X_i^I)$, $i = 1, \ldots, n_I$.

The parametric density estimator based on the external data set offers prior knowledge from the general industry. The correction of this prior knowledge is based on the transformation density estimation of the internal data points that was presented in Chapter 3. This provides us with a consistent and fully nonparametric estimator of the density of the internal data.

The approach is therefore fully nonparametric at the same time as it is informed by prior information based on external data sources. The mixing transformation kernel density estimator has asymptotic behavior as the ones described in Section 3.3.1.

Vector $(\widehat{\alpha}^E, \widehat{M}^E, \widehat{c}^E)$ is a squared-root-n-consistent estimator of the unknown true parameters (α, M, c) minimizing the Kullback–Leibler distance between the generalized Champernowne distribution based on (α, M, c) and the true underlying density of the external data.

Let

$$\hat{l}(T_{\alpha,M,c}(x)) = \frac{1}{n_I}\sum_{i=1}^{n_I} K_b\left(T_{\alpha,M,c}(X_i^I) - T_{\alpha,M,c}(x)\right).$$

Replacing (α, M, c) by the parameter estimates obtained from the external data $(\widehat{\alpha}^E, \widehat{M}^E, \widehat{c}^E)$, it follows that

$$\hat{l}\left(T_{\widehat{\alpha}^E,\widehat{M}^E,\widehat{c}^E}(x)\right) = \frac{1}{n_I}\sum_{i=1}^{n_I} K_b\left(T_{\widehat{\alpha}^E,\widehat{M}^E,\widehat{c}^E}(X_i^I) - T_{\widehat{\alpha}^E,\widehat{M}^E,\widehat{c}^E}(x)\right).$$

Then, the proposed mixing model is expressed as

$$\hat{f}(x, \widehat{\alpha}^E, \widehat{M}^E, \widehat{c}^E) = \frac{T'_{\widehat{\alpha}^E,\widehat{M}^E,\widehat{c}^E}(x)}{\alpha_{01}} \cdot \hat{l}(T_{\widehat{\alpha}^E,\widehat{M}^E,\widehat{c}^E}(x)).$$

This last equation links the results with the intuitive approach discussed in the previous. In this case, $T'_{\widehat{\alpha}^E,\widehat{M}^E,\widehat{c}^E}(\cdot)$ represents the prior knowledge that is obtained from external data by estimating the density of a generalized Champernowne distribution. The term called $\hat{l}(T_{\widehat{\alpha}^E,\widehat{M}^E,\widehat{c}^E}(\cdot))$ represents the estimation of a shape that is close to a uniform distribution if external and internal behaviors are similar. In terms of the notation in the previous section, $T'_{\widehat{\alpha}^E,\widehat{M}^E,\widehat{c}^E}(\cdot)$ corresponds to prior knowledge h_θ, and $\hat{l}(T_{\widehat{\alpha}^E,\widehat{M}^E,\widehat{c}^E}(\cdot))$ is the nonparametric estimator of g_θ. The bandwidth parameter b in the equation is estimated using the approach described in Section 3.4. Note that b is estimated on the transformed internal sample.

Another slightly different mixing model would be to use only partial information from the external data source. The estimator is the same as (4.3) but, instead of using the *prior* knowledge for the parameter vector (α, M, c), *prior* knowledge is only provided by the tail parameter estimators $\hat{\alpha}^E$ and the shift parameter estimator $\hat{c}^E$, while the mode is represented by the estimator $\hat{M}^I$ that is obtained from internal information. With this model we enclose the external parameters $\hat{\alpha}^E$ and $\hat{c}^E$ with a body described internally by $\hat{M}^I$.

4.4 Data Study

This section demonstrates the practical relevance of the just-presented mixing model. We estimate a loss distribution and calculate next year's operational

Table 4.1 *Statistics for event risk category Employment practices and workplace safety*

	N	Maximum	Mean	Median	Deviation	T
Internal data	120	16.00	1.86	0.30	3.66	2
External data	6526	561.43	1.82	0.18	13.41	8

risk exposure by using the common risk measures Value-at-Risk (VaR) and Tail-Value-at-Risk (TVaR) for different levels of risk tolerance. For the severity estimator we employ the transformation estimation using only internal data, the mixing approach expressed in (4.3), and this is benchmarked with the parametric Weibull distribution. For the frequency model we assume that $N(t)$, the number of losses in tear t, is an independent homogeneous Poisson process, $N(t) \in Po(\lambda t)$, with positive intensity λ. By using Monte Carlo simulation with the severity and frequency assumptions we could create a simulated one-year total operational risk loss distribution.

We denote the internal collected losses from the event risk category *Employment Practices and Workplace Safety* with $X_1^I, X_2^I, ..., X_{N^I(t)}^I$ where $N^I(t)$ describes random number of internal losses over a fixed time period $t = 0, ..., T$ and with $N^I(0) = 0$. Further, $X_1^E, X_2^E, ..., X_{N^E(t)}^E$ represent external data from the same event risk category with random number of external losses $N^E(t)$ with $N^E(0) = 0$. Table 4.1 reports summary statistics on each data set.

The mean and median are similar for the two data sets. However, the number of losses, the maximum loss, the standard deviation, and the collection period are widely different. We condition on $N^I(1) = n^I$ and sample the Poisson process of internal events that occur over a one-year time horizon. The maximum likelihood estimator of the annual intensity of internal losses $n^I = 120$ is $\hat{\lambda} = n^I/T = 120/2 = 60$ and denote the annual simulated frequencies by $\hat{\lambda}_r$ with $r = 1, ..., R$ and number of simulations $R = 10,000$. For each $\hat{\lambda}_r$ we draw randomly uniform distributed samples and combine these with loss sizes taken from the inverse estimated cdf of the severity distribution. The outcome is the annual total loss distribution denoted by $S_r = \sum_{k=1}^{\hat{\lambda}_r} \hat{F}^{\leftarrow}(u_{rk}, \theta)$, $r = 1, ..., R$ with $u_{rk} \in U(0,1)$ for $k = 1, ..., \hat{\lambda}_r$, and $\hat{F}^{\leftarrow}(u_{rk}, \theta)$ is obtained from the density that has either been fixed or estimated from the previous methods.

For simplicity, we introduce the abbreviations M_1 for the semiparametric model that only uses internal data and the method described in Chapter 3. M_2 and M_3 are based on (4.3) with parameters estimated from the external data for model M_2 and only partially for model M_3 as described at the end of Section 4.3. M_4 is the purely external data version of the method described in Chapter 3 where the *prior* knowledge is locally corrected with external data so that internal data are not used at all. This corresponds to the situation when a

Table 4.2 *Statistics of simulated loss distributions for event risk category Employment practices and workplace safety*

Model	Mean	Median	Standard dev.
M_1	226	178	149
M_2	244	186	167
M_3	296	213	241
M_4	316	274	322
M_5	100	97	27
Model	$\text{VaR}_{95\%}$	$\text{VaR}_{95\%}$	$\text{VaR}_{95\%}$
M_1	547	765	980
M_2	609	881	1227
M_3	633	936	1331
M_4	801	1165	1644
M_5	249	272	294
Model	$\text{TVaR}_{95\%}$	$\text{TVaR}_{95\%}$	$\text{TVaR}_{95\%}$
M_1	691	889	1180
M_2	787	1025	1385
M_3	820	1143	1487
M_4	1073	1388	1858
M_5	363	383	405

company has not started to collect internal data and must rely entirely on *prior* knowledge. Finally, M_5 is the benchmark model with a Weibull assumption on the severity estimated with internal data. Then, a loss distribution could be calculated with relevant return periods and thereby identify the capital to be held to protect against unexpected losses. The two risk measures used are the VaR

$$\text{VaR}_{\alpha}(S_r) = \sup\{s \in \mathbb{R} \mid \mathbb{P}(S_r \leq s) \leq \alpha\}$$

and TVaR that give us the expectation of the area above VaR and is defined as

$$\text{TVaR}_{\alpha}(S_r) = \mathbb{E}[S_r \mid S_r \geq VaR_{\alpha}(S_r)]$$

for risk tolerance α. Table 4.2 collects summary statistics for the simulated total loss distribution across each model. Among the usual summary statistics we report VaR and TVaR for risk tolerance $\alpha = \{0.95, 0.99, 0.999\}$.

One can see from Table 4.2 that all loss distributions are right skewed since all means are larger than its respective median. Interpreting the VaR results, one notices that the fully *prior* knowledge model M_4 shows much larger values than the others. The benchmark model M_5 predicts the lowest values of the considered models, and if we compare M_5 against the purely internal estimator M_1, model M_5 suggests 30%-40% capital requirement of the total amount M_1 recommends. For the mixing model M_2 we have only 8% higher sample mean and 10%-25% higher VaR values than M_1, and this is due to *prior* knowledge based on external data. If we view the TVaR results, a similar pattern as in VaR could be visualized. The conclusion is that the two mixing models, M_2 and M_3, seem to be stabilizing the estimation process.

4.5 Further Reading and Bibliographic Notes

The proposed mixing model is based on a semiparametric estimator of the loss distribution. The books referenced in [56], [60], [18], and [62] are important introductions to actuarial estimation techniques of purely parametric loss distributions. Cruz [22] describes a number of parametric distributions that can be used for modeling loss distributions. References [31] and [29] focus on the tail and Extreme Value Theory (EVT). Further, the recent paper by Degen et al. [27] discusses some fundamental properties of the g-and-h distribution and how it is linked to the well-documented EVT-based methodology. However, if one focuses on the excess function, the link to the Generalized Pareto Distribution (GPD) has an extremely slow convergency rate, and capital estimation for small level of risk tolerance using EVT may lead to inaccurate results if the parametric g-and-h distribution is chosen as the model. Also, [2] and [28] stress the weaknesses of EVT when it comes to real data analysis. This problem may be solved by nonparametric and semiparametric procedures. For a review of modern kernel smoothing techniques see [73]. During the last decade, a class of semiparametric methods was developed and designed to work better than a purely nonparametric approach (see [4], [50], and [20]). They showed that nonparametric estimators could substantially be improved in a transformation process, and they offer several alternatives for the transformation itself.

The paper by Buch-Larsen et al. [10] maps the original data within $[0,1]$ via the parametric start and correct nonparametrically for possible misspecification. A flexible parametric start helps to mitigate an accurate estimation. Reference [10] generalized the Champernowne distribution (GCD) (see [14] and [15]), and used the cumulative distribution as transformation function. In the spirit of [10], many remedies have been proposed. Gustaffson et al. [46] investigate the performance between symmetric and asymmetric kernels in $[0,1]$, in [48] they estimate operational risk losses with a continuous credibility model, and in [43] and the extended version by [7] the authors have introduced the concept of underreporting in operational risk quantification. Bolancé et al. [5] develop a transformation kernel density estimator by using a double transformation to estimate heavy-tailed distributions. Before describing the new method, we mention other recent methods in the literature. Three papers, [69], [11], and [59], combine loss data with scenario analysis information via Bayesian inference for the assessment of operational risk, and Verrall et al. [72] examine Bayesian networks for operational risk assessment. Reference [35] develops a method that can estimate the internal data distribution with help from truncated external data originating from the exact same underlying distribution. In our approach we allow the external distribution to be different from the internal distribution, and our transformation approach has the consequence that the internal estimation process is guided by the more stable estimator originating from the external data. If the underlying distribution of the external data is close to the underlying distribution of the internal data, then this approach improves the efficiency of estimation. We also present a variation of this approach where we correct for the estimated median such that only the shape of the underlying internal data has to be close to the shape of the underlying external data for our procedure to work well. See also [42].

For the inspiration in the slightly different mixing model, see [48].

We should also cite [19], where there is a reference on combining probability distributions from experts in risk analysis in general. For an excellent presentation about relationships using copulas, see [38].

Chapter 5

Underreporting

5.1 Introduction

Underreporting means that not all operational risk claims in the company are reported. An underreporting function encodes the likelihood that loss of a particular size is being reported. Because in general the probability of reporting increases with the size of the operational risk claim, the density of the observed losses in the reported data set is more heavy tailed compared to the density of all operational risk claims.

In this chapter we set up a model that first defines all the operational risk claims that have occurred — even though not all of them have been reported — and then models the statistical relationship between the actually reported claims and the total number of occurred claims.

An approach to dealing with underreporting has been described by several authors who regard the problem as being one in which the publicly reported data is subject to an unknown (and varying) lower reporting threshold.

We propose an alternative, simpler approach, drawing on the input of risk experts' opinion in the estimation of underreporting to derive estimators of the underlying true publicly reported loss data size distribution. In our example, subject matter experts within an internationally active insurance company were asked to provide percentage estimates of reporting likelihood across the Basel operational risk categories at different loss sizes. These factors are then used to derive an underreporting function that can be combined with the publicly reported data to give an estimator of the underlying true loss distribution. Requirements that apply to the underreporting function are that it must pass through certain points (reflecting the responses given by the subject matter experts), that it should be a continuous function, and that it should be mathematically well behaved beyond the final reporting point. The mathematical approach applied in this situation is termed spline theory, which seeks to interpolate values between the reported values (threshold points), and beyond the final reporting value, extrapolation is applied. More precisely, we utilize a cubic spline approximation, fitting piecewise cubic polynomials between the fixed report levels (termed knots in spline theory). As a result, the approach provides an indication on the true size of the losses. The same function is also involved in the frequency setup, but here in a slightly different form.

The approach is to first formulate the underreporting functions on each event risk category based on the judgment of the risk experts' opinion on the likelihood of reporting losses. Thereafter, a distribution is estimated for the available data combined with the underreporting information to produce an estimate of the true loss distribution that counteracts the impact of underreporting. For illustration, we present an application utilizing publicly reported loss data and introduce different true distributions as benchmarks. We compare the different loss distributions with and without the underreporting effect and evaluate the consequences for operational risk capital requirements in the different scenarios.

5.2 The Underreporting Function

When modeling a publicly reported database, one needs to fit the operational risk exposure to a company's internal situation. This is done here by introducing a function into the model, which could be thought of as a scaling function explaining the likelihood of a loss being reported at different threshold levels. A very simple interpretation of how the function affects the distribution is that, in the beginning and main body, it is upweighted to include more probability mass in that domain, while larger losses are downweighted. This can be called an underreporting function.

The underreporting function should, as well as being continuous, pass exactly through the predetermined reporting level values. The requirement is mathematically termed as interpolating the report levels, which is a prediction made through and between the report levels. A further requirement is to have a continuous and well-defined function after the last observed report level. Mathematically, this is defined as extrapolation and is a prediction made beyond the last observed report level. Of course, a prediction based upon extrapolation is potentially less reliable than a prediction between report levels since one must assume that a historical pattern will continue outside the reported value domain.

Spline theory meets these requirements. A cubic spline approximation constructed by fitting piecewise cubic polynomials between the report levels (the prediction beyond is captured linearly) can be used.

An example of an estimated underreporting function is given next, where the Basel operational risk categories were used as a standard basis from which to estimate the underreporting functions. Several subject matter risk experts from a major international insurance company provided estimates of underreporting numbers. This led to percentage estimates of the reporting likelihood at different predetermined loss sizes. The risk event categories are shown in Table 5.1.

Data provided by the expert opinion on the probability of reporting a loss of a given amount for each type of risk have been used to produce the underreporting functions. Figure 5.1 presents the estimated probability of reporting for each risk category. It can be seen that the event risk category number seven

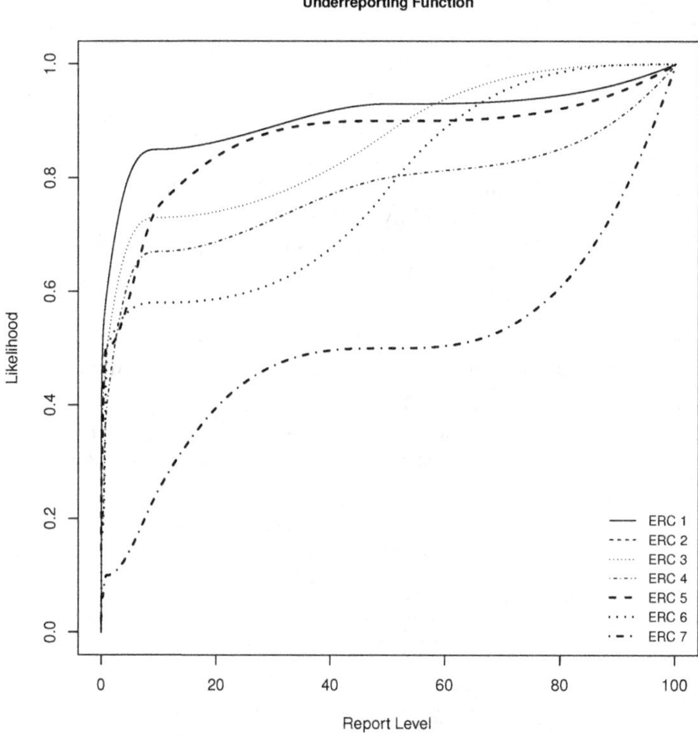

Figure 5.1: *The estimated underreporting function.*

(which corresponds to Execution, Delivery, and Process Management) has the lowest reporting likelihood, which means that losses for this kind of operational risk are likely to be underreported. On the other hand, smaller losses that belong to category number five (which corresponds to Clients, Products, and Systems Failure) have a much higher probability of being reported than small losses for other types of risks. It is also noteworthy that, for all event risk categories, expert opinion assesses that the probability of reporting a loss is smaller or equal than 99%, even for very large losses. In fact, this acts as a bound, and it implies that there is at least a 1% chance that a very large loss occurs and is not reported. The estimation of the underreporting function is essential in our approach, but unfortunately, very little is known about the magnitude of underreporting and even less about the validity of experts' estimation. We think that further research needs to be done on the stability of these estimates, and we believe that here we provide a starting point for discussion.

Table 5.1: *Event risk categories*

Category number	Definition
1	Internal Fraud
2	External Fraud
3	Employment Practices and Workplace Safety
4	Business Disruption
5	Clients, Products, and System Failures
6	Damage to Physical Assets
7	Execution, Delivery, and Process Management

Table 5.2: *Descriptive statistics of the seven event risk categories' reported losses*

Event risk category	Number of losses	Maximum loss	Sample median	Sample mean	Standard deviation
1	1247	6683.8	1.82	32.24	269.43
2	538	910.6	2.14	15.60	69.68
3	721	221.9	1.98	7.84	20.04
4	45	117.6	5.88	22.46	33.25
5	6526	11228.7	3.74	36.59	268.28
6	2395	39546.4	2.35	74.91	1192.55
7	75	104.6	1.56	7.39	17.72

5.3 Publicly Reported Loss Data

The publicly reported loss database used for this section comprises information on over 10,000 events including financial loss suffered, dates, locations, description, and loss category assignment across a range of global organizations. It is subject to quality assurance and continual updating by the vendor. Of course, an organization could compile its database using news reported in the media if possible. There is a considerable variation in both the numbers and sizes of losses reported across categories. In Table 5.2 we give a summary on each event risk categories' statistics.

The second column in Table 5.2 shows the number of observations for each event risk category. There is considerable variation in the number of losses; event risk category 5 and 6 has 6526 and 2395 reported losses, respectively, compared to category 4, which has 45 losses. Column three to six show some empirical results on each event risk category. Note that the mean is significantly larger than the median in all cases, consistent with right-skewed distributions.

As a starting point we use a parametric modeling approach to the operational risk losses available in the public source. Sometime we can call this the external parametric model.

We assume two benchmark parametric distributions (lognormal and

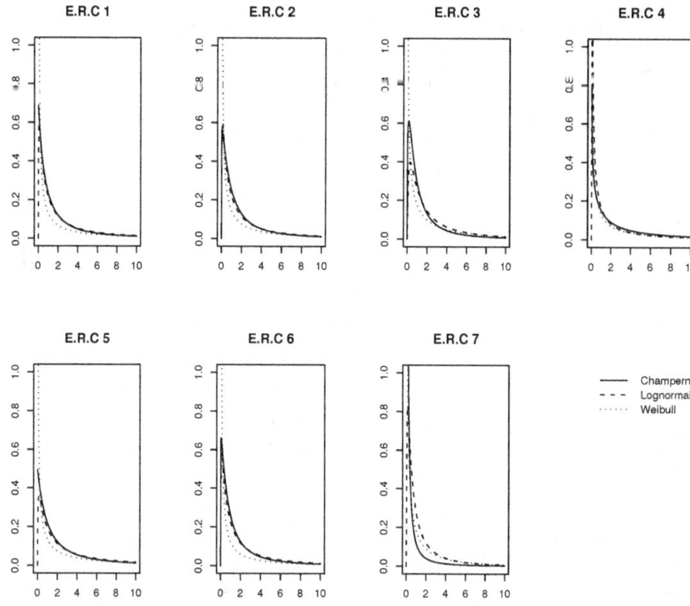

Figure 5.2 *The estimated parametric densities for Event Risk Categories (ERC) 1 to 7.*

Weibull) that are potentially suitable and widely used in the industry for the true losses. We also consider the generalized Champernowne distribution whose characteristics are appealing when working with operational risk losses. The distribution shows a similar shape to the lognormal distribution in the beginning and middle part of the domain and a convergence to the heavy-tail Pareto distribution for very large values as illustrated in Chapter 2.

Figure 5.2 presents the estimated densities on event risk category 1 to 7 under different scenarios. Three different probability distributions were fitted (lognormal, Weibull, and generalized Champernowne).

Figure 5.3 shows on the left the observed density for event risk category number one. There is no correction for underreporting. On the right ("true" density for ERC 1), the estimation accounts for the existing underreporting function are presented. The method to deal with the correction is described in the following section. For the other six event risk categories, a similar shape is expected.

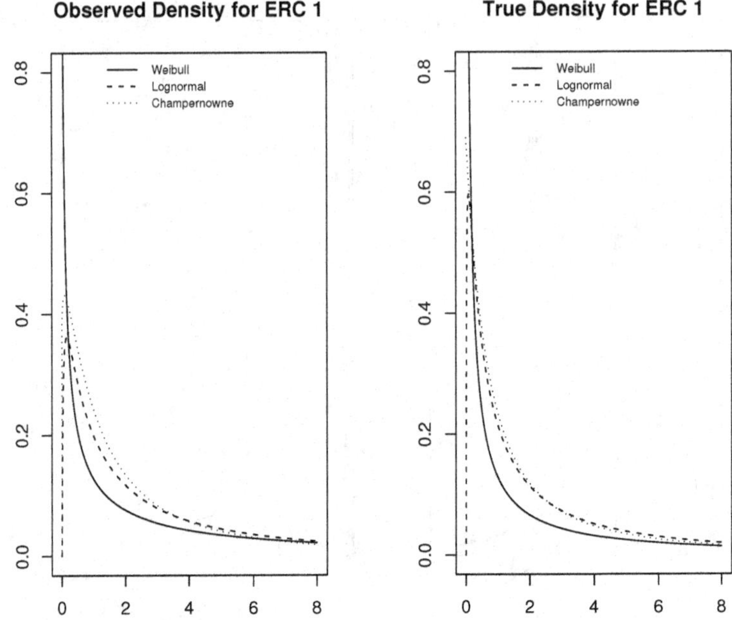

Figure 5.3 *Estimated observed and true densities based on operational risk data originating from the first event risk category.*

5.4 A Semiparametric Approach to Correction for Underreporting

Our correction for underreporting on frequencies assumes that the true number of claims in each category is a random variable that follows a Poisson distribution with intensity λ. The frequency distribution including underreporting is then defined as follows. Let $I(i)$ be a standard indicator function taking the value zero or one if the i-th claim is not observed or observed, respectively. Note that we use the word *observe* but we mean that we only observe a claim if the loss is reported. In fact, a loss may occur, but if it is not reported, then we say that it is not *observed.* The observed number of claims is then defined by the summation of $I(i)$ for all occurred claims. If we let n be the observed number of claims from the publicly reported loss data base, the underreporting is then expressed as an underreporting function $u(x)$ that depends on the size of the claim. Therefore, the underreported frequencies follow a Poisson distribution with intensity λP_u , where P_u is the posterior probability that the claim is reported.

5.4.1 Setting up a Model for the Sampling of Operational Risk Claims with Underreporting

Assume that N_O independent identically distributed (*iid*) operational risk claims, $X_1^O, X_2^O, ..., X_{N_O}^O$, with density f_O have occurred where N_O is a stochastic Poisson(λ)-distributed variable. Since we do not observe all these N_O claims, let $I(i)$ be an indicator function taking the value 1 if the i-th claim is observed and 0 otherwise, and let $I(1), I(2), ..., I(N_O)$ be *iid* stochastic variables indicating reported and not-reported claims. The stochastic variable $N_R = \sum_{i=1}^{N_O} I(i)$ is therefore the reported number of claims. Let $X_1^R, X_2^R, ..., X_{N_R}^R$ be the N_R reported claims from the operational risk data set, and assume that these N_R claims given $N_R = n_R$ are *iid* with density f. We assume furthermore that the underreporting function u only depends on the value of the claim

$$u(x) = P(I(i) = 1 | X_i^O = x), i = 1, ..., N_O.$$

Under this model the probability of observing an operational risk claim can be written as

$$P_{u,f_O} = \int_0^\infty f_O(w)u(w)dw.$$

As a result, the stochastic variable N_R is Poisson distributed with mean $\lambda P_{u,f_O}$. The relationship between the density of the reported operational risk claims and the density of all operational risk claims is

$$f(x) = \frac{f_O(x)u(x)}{P_{u,f_O}}. \tag{5.1}$$

We model (5.1) with a parametric f as an *a priori* model to obtain a model corrected in a nonparametric way. The nonparametric correction is obtained for $N_R = n_R$ by using the fact that, if we transform the reported data set $X_1^R, X_2^R, ..., X_{N_R}^R$ with the cumulative distribution function $F(x) = \int_0^x f(w)dw$, we obtain a data set $Y_i^R = F(X_i^R)$, $i = 1, ..., n_R$ with density h, where h is a uniform density.

A parametric model can be used in (5.1); however, nonparametric density estimates can be used in order not to impose a given shape.

5.4.2 A Transformation Approach to Tail Flattening Accounting for Underreporting

We wish to find an appropriate nonparametric smoothing estimator for the density f_O of our operational risk claims. By doing so, we will be able to adjust appropriately for underreporting by means of nonparametric smoothing. Adjusting is always nontrivial in nonparametric smoothing, and there seem to be many methods to adjust for an underreporting function in a nonparametric way. We have chosen the simplest possible method in terms of implementation and analysis. However, further research might lead to other methods for the nonparametric correction of underreporting.

We have observations with density f and, based on (5.1) above, we express f_O as a function of f and u in the following way:

$$f_O(x) = \frac{f(x)\{u(x)\}^{-1}}{\int_0^\infty f(w)\{u(w)\}^{-1}dw}.$$

We know that f can be estimated by

$$\widehat{f}(x) = \frac{F'(x)}{n_R\alpha_{01}(F(x),b)}\sum_{i=1}^{n_R} K_b\left(F(x) - F(X_i^R)\right),$$

where $K_b(\cdot) = \frac{1}{b}K(\frac{\cdot}{b})$ is a kernel function with bandwidth b. Bandwidth choice has already been discussed in Chapter 3.

An obvious estimator of f_O is therefore

$$\widehat{f_O}(x) = \frac{\widehat{f}(x)\{u(x)\}^{-1}}{\int_0^\infty \widehat{f}(w)\{u(w)\}^{-1}dw}.$$

When we consider the asymptotic properties of this estimator, we notice that the asymptotic distribution of $\widehat{f}$ is well known.

We can derive the asymptotic properties of $\widehat{f_O}$ from the asymptotic properties of $\widehat{f}$. Let

$$A = f(x)\{u(x)\}^{-1},$$

$$\widehat{A} = \widehat{f}(x)\{u(x)\}^{-1},$$

$$B = \int_0^\infty f(w)\{u(w)\}^{-1}dw,$$

and

$$\widehat{B} = \int_0^\infty \widehat{f}(w)\{u(w)\}^{-1}dw.$$

Then

$$\widehat{f_O}(x) - f_O(x) = \widehat{A}\widehat{B}^{-1} - AB^{-1} = \widehat{B}^{-1}\left(\widehat{A} - A\right) - A\widehat{B}^{-1}B^{-1}\left(\widehat{B} - B\right).$$

Therefore, $\widehat{f_O}(x) - f_O(x)$ is equivalent from an asymptotic point of view to

$$B^{-1}\left(\widehat{A} - A\right) - AB^{-2}\left(\widehat{B} - B\right).$$

Based on this quick ordering of terms, we can write up the asymptotic theory of $\widehat{f_O}(x)$. We omit the proof, which is based on the Theorem 1 in Chapter 3, and the fact that the variance of $\widehat{B}$ is of a lower order of magnitude due to the integration, while the integrated bias of $\widehat{B}$ still is of the original order b^2. From this theoretical result, we obtain two main conclusions for our adjustment for undersmoothing. First, the bias is affected by the way the adjustment is carried out. Second, the standard deviation of the final estimator is increased by a local element, the square-root of the underreporting function and by a global element, the square-root of an average of the underreporting function. One can also verify that, in the situation where the underreporting function is incorrect, the adjustment method simply results in a biased estimator where f_O is different from the function we would like to obtain. However, the theoretical results below are still valid in that situation.

Theorem 3 *Let the transformation function F and the underreporting function u be two times differentiable known functions. Assume that f_O is also two times continuously differentiable. Then the bias of $\widehat{f_O}$ is*

$$\mathbb{E}\left[\widehat{f_O}(x)\right] - f_O(x) = \alpha_{21} b^2 \left[B^{-1} B_x \{u(x)\}^{-1} - AB^{-2} \int B_w \{u(w)\}^{-1} dw\right] + o(b^2),$$

where $B_x = \left(\left(\frac{f(x)}{T'(x)}\right)' \frac{1}{T'(x)}\right)'$, *and the variance is given by*

$$\mathbb{V}\left[\widehat{f_O}(x)\right] = \{u(x)B\}^{-1} (n_O b)^{-1} \alpha_{02} F'(x) f(x) + o\left\{(n_O b)^{-1}\right\}.$$

More details and the proofs can be found in the articles and books mentioned in the last section.

5.5 An Application to Evaluate Operational Risk with Correction for Underreporting in Publicly Reported Losses

In this section we simulate an operational risk loss distribution when the correction for underreporting is done for every event risk category of publicly reported losses. This is obtained using a severity distribution for event risk category number one and correspondingly for all the other events) together with a Poisson-based frequency through 10,000 draws for one-year operational loss. Summation of amounts from the seven event risk categories provides a single value for the simulated distribution. For the severity distributions we have six different model assumptions, three distributions with and without the underreporting correction effect. For the frequency we have four different scenarios, the observed and the corrected frequencies for each probability distribution. The different frequencies for the different event risk categories can be found in Table 5.2.

Figure 5.4 shows six histograms of operational risk exposure. The top row is total loss distributions without an underreporting effect. The second row includes the underreporting effect. Note that the generalized

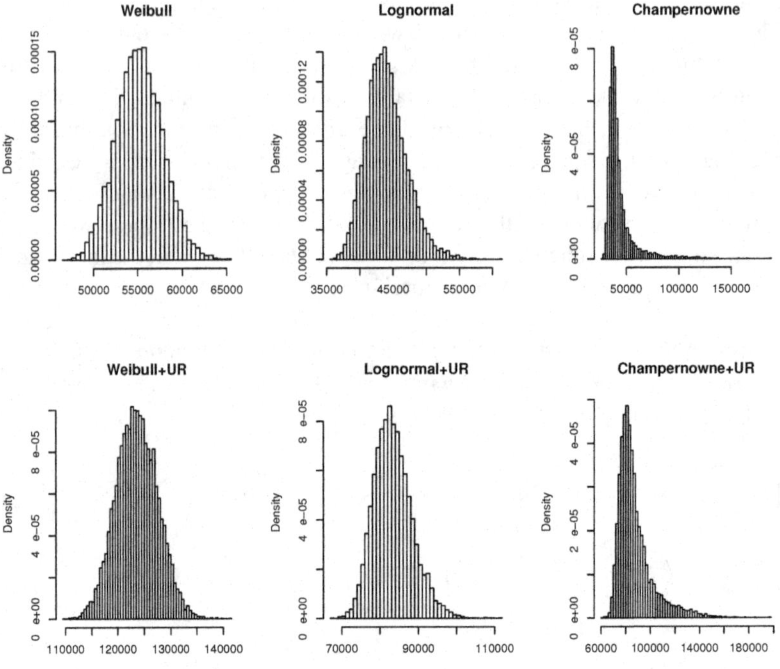

Figure 5.4 *Density for event risk category number one (direct versus corrected for underreporting).*

Champernowne model, both with and without underreporting, presents a heavy-tail characteristic.

Table 5.3 shows the Value-at-Risk (VaR) and Tail Value at Risk (TVaR) under the tolerance level for the three specific models when we account for underreporting as estimated by the experts. Table 5.4 shows what happens if it is assumed that there is no underreporting. For increasing values of tolerance, the VaR and TVaR increase is expected. It is also shown that, when underreporting is neglected, two of the three models have a similar behavior. Only the generalized Champernowne distribution, which has a heavy-tailed behavior, would imply larger capital requirements, especially for the 0.999 tolerance level. Underreporting brings effects to capital requirements calculation. The greater reporting of lower value losses will lead to estimated distributions that could have lighter-tailed characteristics and hence lower capital estimates. On the other hand, where there is underreporting at larger loss values, the effect

Table 5.3: *Statistics for the loss distribution with a correction for underreporting*

	Risk Measure					
Model	VaR 0.95	VaR 0.99	VaR 0.999	TVaR 0.95	TVaR 0.99	TVaR 0.999
Champernowne	439262	496651	580983	475650	535491	636321
Lognormal	136252	140715	146987	139301	144499	157573
Weibull	126701	129317	132121	128451	131250	139782

Table 5.4: *Statistics for the loss distribution without a correction for underreporting*

	Risk Measure					
Model	VaR 0.95	VaR 0.99	VaR 0.999	TVaR 0.95	TVaR 0.99	TVaR 0.999
Champernowne	69034	81202	95745	77003	89809	123468
Lognormal	63057	65804	69189	65021	68133	80326
Weibull	68809	71318	73239	70438	72597	77669

will be to increase the required capital. In our study, it appears that the latter effect outweighs the former; however, the impact in company-specific situations will vary with the specific data set being considered and the degree of underreporting estimated. Moreover, it will depend on the input from subject matter expert. Interestingly, the generalized Champernowne model differs significantly from the other two models because the values for VaR and TVaR are as much as twice the size of the ones obtained for the other two models.

We have addressed the question of underreporting, which is equivalent to quantifying operational risk when there are two sources of information. One comes from data on the frequency and amount of losses, which may be internal or external. The other source of information is based on the risk experts' opinion on the likelihood of reporting losses with respect to the loss amounts. We have used professionals to agree on an estimate of the probability of reporting a loss for several risk event categories. We observe that the profiles vary significantly for one type of risk to the other. As expected, large amounts are more likely to be reported than small amounts.

We have then looked at the risk evaluation of the sum of risks and have confirmed that the use of a distribution like the generalized Champernowne distribution that is heavy tailed substantially increases the estimated capital requirements under the conditions of our study. On the other hand, we have been able to assess that ignoring the underreporting phenomenon can lead to a biased estimate of the VaR and TVaR. The approach described here is general in the sense that it could be applied within any organization. In our study we have found that the overall effect of considering underreporting is to increase the capital requirement. However, this finding may well not be universal since two opposing effects are in operation and could lead to a lower capital requirement. The overall effect will depend on both the expert judgment on reporting and the external losses chosen for modeling. Typically, though, we expect higher capital requirements to result from risks taken in all the losses, in the calculation reported as well as not reported.

We have offered a way of evaluating the difference between the capital needs to cover operational risk in practice and the capital standards obtained from data that experts believe to be biased. We call for the further consideration of such methods and also for deeper studies on the underreporting phenomenon.

5.6 An Application to Evaluate Internal Operational Risk with Correction for Underreporting

We use again a publicly available database with more than 5000 financial operational risk events from a range of global organizations. For each operational risk event, we have information on date, location, loss category, and a description of the event. The reported operational risk events are categorized into six different event risk categories. These categories correspond to exactly the same used in the previous section, but event risk category five has been removed. Therefore, the following categories will be studied:

1. Internal fraud
2. External fraud
3. Employment practices and workplace safety
4. Business disruption
5. Damage to physical assets
6. Execution, delivery, and process management

As seen in Table 5.5 and as in the previous sections, the number and severity of the losses differ considerably in the six categories. This information is similar to the one shown in Table 5.2. Note that the number of losses range from 45 events to 2395, and the loss amounts range from just over 100 million to almost 40 billion. As with most operational risk data sets, the mean is significantly larger than the median, which indicates that the distribution of operational risk events is right skewed.

The last column in Table 5.5 provides information for the internal operational risk model. Since the external database does not provide a reliable estimate of the annual frequency of each event risk category, these numbers are

Table 5.5: *Number of reported operational risk losses in our external database*

ERC	Number of losses	Maximum loss	Sample median	Sample mean	Standard deviation	Annual frequency
1	1247	6683.8	1.82	32.24	269.43	10
2	538	910.6	2.14	15.60	69.68	20
3	721	221.9	1.98	7.84	20.04	28
4	45	117.6	5.88	22.46	33.25	11
5	2395	39546.4	2.35	74.91	1192.55	3
6	75	104.6	1.56	7.39	17.72	52

estimated using scenario analysis. Scenario analysis is based on the presumed risk of a given insurance company, and we will assume that these are the frequencies that have been reported for one year in a company or estimated as averages of several periods.

We use the six corresponding underreporting functions proposed in Section 5.2 on the basis of expert judgments(see Figure 5.1 for details). These underreporting functions are based on parametric modeling and a lot of aggregated experience, and therefore can be seen as known functions with respect to our asymptotic theory in Section 5.4. In the same way, we assume that our parametrically defined transformation function is known. This type of argument is well known in semiparametric density estimators like ours. We consider the three parametric models and estimate parameters by maximum likelihood. The transformations we use are the parametric cumulative distribution functions (cdfs) produced by integrating the following parametric densities for event risk category j, which have been introduced in Chapter 2.

- The generalized Champernowne probability distribution function (pdf):

$$f_{\theta_j}(x) = \frac{\alpha_j (x+c_j)^{\alpha_j - 1}\left((M_j+c_j)^{\alpha_j} - c_j^{\alpha_j}\right)}{\left((x+c_j)^{\alpha_j} + (M_j+c_j)^{\alpha_j} - 2c_j^{\alpha_j}\right)^2}, \tag{5.2}$$

  where $\theta_j = \{\alpha_j, M_j, c_j\}$, and j denotes the risk category.

- The lognormal pdf:

$$f_{\eta_j}(x) = \frac{e^{-\frac{1}{2}\left(\frac{\log x - \mu_j}{\sigma_j}\right)^2}}{x\sigma_j\sqrt{2\pi}}, \tag{5.3}$$

  where $\eta_j = \{\mu_j, \sigma_j\}$.

- The Weibull pdf:

$$f_{\varsigma_j}(x) = \frac{\gamma_j}{\beta_j}\left(\frac{x}{\beta_j}\right)^{\gamma_j - 1} e^{-\left(\frac{x}{\beta_j}\right)^{\gamma_j}}, \tag{5.4}$$

where $\varsigma_j = \{\gamma_j, \beta_j\}$.

5.6.1 *Aggregated Analysis Incorporating six Event Risk Categories*

We use Monte Carlo simulation to calculate the 99.5 VaR and TVaR for our various versions of estimated distributions. The VaR measure gives us insight into expected maximal losses for risk tolerance $\alpha = 0.995$.

The VaR measure is a common risk measure. However, in contrast to TVaR, VaR is not a coherent risk measure. That means that VaR does not always fulfill the important property of subadditivity, which, loosely speaking, means that one will never benefit from splitting up a risk.

Our chosen values of α are inspired by Basel II, which specifies standards within the advanced measurement approach.

When simulating, we draw 10,000 operational claims numbers for each risk category using the frequencies from our scenario analysis as our Poisson parameters (see the last column of Table 5.5).

$$r_{i,j} \sim Po\left(\lambda_j\right), \quad i = 1,2,...,10,000 \; j = 1,2,...,6.$$

Then, for each of the 60,000 simulated number of operational risk claims, the $r_{i,j}$s, we draw $r_{i,j}$ independent identically distributed stochastic variables by means of our nonparametric estimators of the distributions of operational risk claims. First we sample $r_{i,j}$ uniform distributions:

$$u_{i,j,k} \sim U\left(0,1\right), \quad k = 1,2,...,r_{i,j}.$$

We then calculate the simulated aggregated claim amount for the jth risk category and ith simulation:

$$x_{i,j} = \sum_{k=1}^{r_{i,j}} \widehat{F}_i^{-1}\left(u_{i,j,k}\right)$$

where $\widehat{F}_j^{-1}$ is the inverse of the estimated cumulative distribution function $\widehat{F}_j$ for the jth risk category. Our risk measure is then based on the 10,000 values of

$$x_{i,\cdot} = \sum_{j=1}^{6} x_{i,j}.$$

5.6.2 *Results*

Table 5.6 presents the corrected frequencies for each risk category. The annual frequencies for the company are presented in the first row. The six following rows give the corrected frequencies based on different distribution assumptions with adjustment for the underreporting effect. The abbreviation Ch.UR

Table 5.6 *Reported frequency for each risk category and corrected risk frequencies after adjusting for underreporting with and without nonparametric correction*

	Risk category					
	1	2	3	4	5	6
Unadjusted	10.0	20.0	28.0	11.0	3.0	52.0
Ch.UR	14.4	28.2	46.8	17.9	5.7	290.6
Ch.UR.KS	13.7	27.6	44.9	18.2	5.4	270.4
Ln.UR	14.1	28.1	45.3	17.6	5.6	251.7
Ln.UR.KS	14.1	29.1	48.1	22.7	5.3	391.3
We.UR	13.6	28.6	47.9	22.4	5.3	386.6
We.UR.KS	14.2	33.7	49.3	24.7	5.6	400.3

is the generalized Champerknowne distribution adjusted for underreporting, Ch.UR.KS is the generalized Champerknowne distribution adjusted for underreporting and kernel smoothing.

Table 5.7 presents the total operational loss of the institution; the mean, median, standard deviation, and the 99.5% VaR and TVaR based on different underlying distributions. We consider the three parametric models with and without correction for underreporting, and with and without the nonparametric correction based on kernel smoothing. We normalize all results by the results obtained by using the parametric Weibull distribution without any kind of correction, which seems to be the most popular model among practitioners at the moment.

From Table 5.7 we see that incorporating underreporting clearly increases both the mean value, the median value, the standard deviation, the 99.5% quantile, and the 99.5% TVaR. This is because more claims are being incorporated into our model when underreporting is taken into account. However, most of these claims are small, and therefore, the 99.5% quantile is less affected by taking underreporting into account than the other measures in Table 5.7.

When underreporting is not accounted for, kernel smoothing has a tendency to correct the tail into a heavier tail that increases most of the considered measures of risk, while kernel smoothing has the opposite effect when we account for underreporting. It seems that, when underreporting is present, a major correction is necessary in order to have a sufficient small claim mass in the distribution. This correction takes mass from the tail of the distribution and moves it to the smaller values of the distribution. It is also clear from Table 5.6 that, while different parametric models give very different answers, our kernel-smoothed correction has a stabilizing effect; it is clear that this stabilizing effect affects the quantile estimation as well as the Tail-Value-at-Risk

Table 5.7 *Statistical data for the total loss amount normalized by the unadjusted Weibull distribution*

	Mean	Sd	Median	VaR 99.5%	TVaR 99.5%
We	1	1	1	1	1
We.KS	1.71	2.09	1.71	1.88	1.96
We.UR	2.30	1.56	2.34	1.93	1.90
We.UR.KS	1.92	2.21	1.94	2.02	2.11
Ln	0.98	1.54	0.90	1.35	1.45
Ln.KS	1.64	2.26	1.63	1.83	1.84
Ln.UR	2.28	3.02	2.18	2.84	2.90
Ln.UR.KS	1.87	2.43	1.81	1.99	2.05
Ch	1.12	3.11	0.86	2.51	2.81
Ch.KS	1.56	2.14	1.54	1.75	2.14
Ch.UR	2.50	4.65	2.21	3.85	4.10
Ch.UR.KS	2.12	2.74	2.01	2.11	2.36

estimation. It does not really matter very much which of the three parametric models we use for our pilot study when a stabilizing kernel smoothed correction is performed. However, the choice of parametric model seems to be crucially important if one decides to stick to a purely parametric approach. One can, for example, conclude that one gets estimates of the exposure to operational risk that are too optimistic if one uses the widely used parametric Weibull distribution without correcting for underreporting and without a nonparametric correction based on kernel smoothing. We therefore recommend that regulators and practitioners start looking for other approaches with more realistic estimates of the tail behavior of actuarial loss distributions.

5.7 Further Reading and Bibliographic Notes

Reference [47] addressed further details on underreporting. See also [36] and [3].

For an updated overview about quantifying operational risk in a general insurance company, see [71] and the discussion therein.

A deeper treatment of spline approximation can be found in [34] and [25].

In [54] there is an extensive discussion of the difference between adjusting for the design internally (inside the integral) or externally (outside the integral) in the standard nonparametric estimation problem. While [10] describes the details behind a standard correction based on the simple boundary correction procedure of density estimation, see also [70] or [73].

Chapter 6

Combining Underreported Internal and External Data for Operational Risk Measurement

6.1 Introduction

In this chapter we combine the techniques introduced in the previous chapters for operational risk: the data are underreported, they are truncated and, besides that, external data are often needed to incorporate more information [1].

In order to show the methodology in an intuitive way, we first exemplify every calculation with simulated data. In the first step, the simulated internal data set and the simulated external data set have not been truncated, and all operational risk losses can be assumed to be reported. Then, we go through the analysis of these data sets separately. Afterwards, we introduce into our simulation the underreporting and truncation of the internal and external data sets, and then we show how to overcome the consequences of these sampling restrictions in the statistical analysis. In section 6.4 we combine internal and external data, and we introduce a nonparametric correction. Finally, we discuss an empirical application to real data.

Many different issues arise in practice when insurers analyze their operational risk data. We will focus on three of them:

1. Internal information is scarce.
2. External data can be shared by all insurers under the auspices of a consortium, but pooling the data requires a mixing model that combines internal and external information.
3. Data used to assess operational risk are sometimes underreported, which means that losses do happen that are unknown or hidden. Data on operational losses are sometimes unrecorded below a given threshold due to existing habits or organizational reasons, which means that information on losses below a given level is not collected. One should not confuse underreporting with unrecording because the latter is a voluntary action not to collect all those losses below a given level.

[1]We recommend the reader to go through the guided examples provided in Chapter 7 before addressing the more sophisticated approach presented here, where several methods are applied simultaneously

When estimating an internal operational risk loss distribution, a major obstacle is that not all losses are observed. Of course, the prior knowledge incorporated will ensure exposure is present in some intervals; while these losses are in general large in size, some intervals will be underreported in the loss distribution modeling. To estimate such an underreporting function from the internal data (and/or the external data) requires a complicated mathematical procedure.

In this chapter we address the quantification of operational risk when accounting for the three problems indicated above. In fact, methodologies have already been proposed before for every subproblem. Here we present a unified approach that can cope with all of them at the same time. The most difficult part is trying to find ways to combine information from several sources (i.e., internal and external data), each one coming from institutions having different collection thresholds and possibly different reporting behaviors. We also aim to present models that correct for underreporting and combines internal and external data. We merge the theory developed in the previous chapters and generalize the methodology incorporate collection thresholds as well. We proceed as follows: Section 6.2 lays out data availability and discusses the issues a company could visualize when sophisticated methodologies should be utilized for operational risk assessment. In Section 6.3 the underreporting model is presented. Section 6.4 presents how two sources of data can be utilized and an unrecording problem exists. The proposed model is extended to allow for underreporting both for the internal and external data. In Section 6.5, an application on real operational risk data is provided that illustrates the interpretation and usefulness of the proposed estimation framework. For each model, the total loss distributions with holding period of one year for specific risk tolerance levels is estimated and its quantiles are derived.

6.2 Data Availability

In general, the collection period of internal operational risk losses is very short for most insurers, which itself generates a scarce sample to estimate and validate on. Therefore, nobody doubts that complementing internal data with more abundant external data is desirable. A suitable framework to overcome the lack of data is getting data from a consortium. Individual insurers come together to form a consortium, where the population from which the data is drawn is assumed to be similar for every member of the group. We know that this can be argued because not all insurers are the same, and one can raise questions about whether consortium data can be considered to have been generated by the same process, or whether pooled data do reflect the size of each insurer transaction volume. We will avoid this discussion here and assume that scaling has already been corrected for.

Let $X_1^I, X_2^I, \ldots X_{n_I}^I$ be occurred internal losses from a specific event risk category with total occurred number of losses denoted by n_I over some time period. Let $X_1^E, X_2^E, ...X_{n_E}^E$ be n_E occurred external losses (e.g., consortium data or publicly reported losses) from the same event risk category over a similar time period.

To illustrate our presentation, we assume that the internal and external losses are lognormally distributed with the same location parameter and different scale parameter. More precisely, we have assumed X_i^E is lognormally distributed with $\mathbb{E}(\log X_i^E) = 0$ and $\mathbb{V}(\log X_i^E) = 3/2$, and X_i^I is lognormally distributed with $\mathbb{E}(\log X_i^I) = 0$ and $\mathbb{V}(\log X_i^I) = 5/2$. In our example, we let $n_I = 50$ and $n_E = 500$, and usually the relation $n_I < n_E$ holds. The assumption of similar location parameter between the internal and external data sources is based on the assumption that similarity exists between members of the consortium. In our simulated sample, the maximum internal occurred loss is 96 and the maximum loss for the external sample turned out to be 919.

Figure 6.1 shows the simulated data sets for the internal and external distribution. Initially, we will consider all occurred losses, but we have used different symbols to identify the cases that will not be included in the truncated sample and later in the undererporting sample. All losses below the threshold (denoted by the crosses) are not reported at all, so they will disappear in the truncated samples. The underreported losses over the collection threshold are marked with grey circles. There are 4 internal losses and 14 external losses that are not reported, even though their size exceeds the collection threshold. Our samples with underreporting will exclude those cases that are not reported and will work only with the reported losses (dark circles).

6.2.1 *Minimum Collection Threshold*

Companies usually fix a minimum threshold level so that all losses below this level are not reported. Define the internal collection threshold by C_I, and let C_E be the external collection threshold. We assume that consortium data refer to losses above a given C_E level. Then, we define a truncated sample by $\left(X_i^{I,C_I}\right)_{1 \leq i \leq n_{I,C_I}}$ as occurred internal losses above the minimum collection threshold C_I, and $\left(X_i^{E,C_E}\right)_{1 \leq i \leq n_{E,C_E}}$ as occurred external losses above the minimum collection threshold C_E. Here, n_{I,C_I} and n_{E,C_E} are the total number of occurred losses larger than the internal and external thresholds, respectively.

In our illustration we assume that the internal collection threshold $C_I = 1$, and the external collection threshold $C_E = 10$. Then, the occurred total number of losses above the thresholds turns out to be $n_{I,C_I} = 21$, and $n_{E,C_E} = 89$.

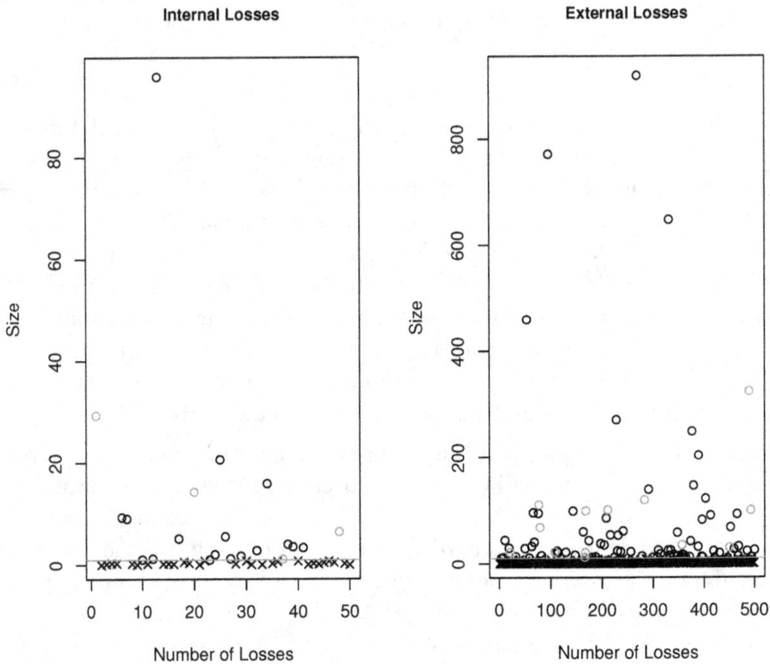

Figure 6.1 *Simulated internal and external loss data from a specific event risk category. The thresholds are represented with grey lines for the two samples, and the losses smaller than the thresholds are marked with crosses. The losses above the threshold that are not being reported are marked with grey circles.*

6.2.2 *Modeling Each Data Set Separately*

Internal, and external data should first be modeled. Here we will use the lognormal distribution for simplicity, but in practice one can make another distributional assumption, either parametric or non-parametric.

If all occurred losses $X_1^I, X_2^I, ..., X_{n_I}^I$ and $X_1^E, X_2^E, ..., X_{n_E}^E$ were available and a lognormal distribution was assumed for the operational risk severity distribution, the target density would be

$$f_{O,d}(x) \quad = \quad \frac{1}{x\sigma_d\sqrt{2\pi}} e^{-\frac{1}{2\sigma_d^2}(\log(x)-\mu_d)^2}, \; \mu_d \in R, \; \sigma_d > 0, \; x \geq 0, \quad (6.1)$$

where d indicates either I for internal data or E for external data.

The location and the scale parameters $\{\mu_d, \sigma_d\}$, $d = I$ or $d = E$, in the lognormal density can easily be estimated on each data set using the maximum likelihood estimation method.

Table 6.1: *Quantile values for several models based on simulated data*

Internal Data				
Model	Sample size	$q = 0.95$	$q = 0.99$	$q = 0.995$
$\hat{F}^{\leftarrow}_{O,I}(x_q)$	100	11.79	32.77	47.64
$\hat{F}^{\leftarrow}_{I}(x_q)$	21	50.46	183.74	299.25
External Data				
Model	Sample size	$q = 0.95$	$q = 0.99$	$q = 0.995$
$\hat{F}^{\leftarrow}_{O,E}(x_q)$	500	61.08	335.58	626.14
$\hat{F}^{\leftarrow}_{E}(x_q)$	89	311.80	1111.95	1811.51

Note: We denote by $\hat{F}^{\leftarrow}_{O,d}(\cdot)$, with $d = I$ or E, the estimated quantile values for the lognormal model (6.1), for internal and external data, respectively. $\hat{F}^{\leftarrow}_{d}(\cdot)$, with $d = I$ or E, denotes estimated quantile values for the truncated lognormal models (6.2) and (6.3), respectively.

When only data above a given threshold are observed, the truncated lognormal density should be used. Its density is defined depending on the truncation level C_I or C_E as

$$f_I(x) = \frac{f_{O,I}(x)}{\int_{C_I}^{\infty} f_{O,I}(w)dw} \text{ for } x \geq C_I \tag{6.2}$$

and

$$f_E(x) = \frac{f_{O,E}(x)}{\int_{C_E}^{\infty} f_{O,E}(w)dw} \text{ for } x \geq C_E, \tag{6.3}$$

where $f_{O,I}(\cdot)$ and $f_{O,E}(\cdot)$ correspond to (6.1) for the internal and the external severity distribution, respectively.

Table 6.1 presents the quantiles with return periods 1 in 20 years loss (95%), 1 in 100 years loss (99%), and 1 in 200 years loss (99.5%), where $\hat{F}^{\leftarrow}_{O,d}(\cdot)$ and $\hat{F}^{\leftarrow}_{d}(\cdot)$, with $d = I$ or E, denote estimated quantile values for the lognormal and the truncated lognormal model, originating from models (6.1), (6.2), and (6.3) when the original and the truncated data are used for estimation, respectively. The large values of the external truncated distribution can be explained by the large probability mass below the threshold C_E, so it shows a big truncation effect and possibly some instability due to a small sample size.

6.3 Underreporting Losses

The concept of underreporting has been introduced in the context of operational risk in the purely parametric case and then extended to a semiparametric version. These developments provided an analysis of the quantitative impact of the failure to report all operational risk losses. When estimating operational risk losses, we know that not all losses are reported. Losses below the collection period will not be recorded at all. Those losses above the collection threshold will mostly be recorded, but sometimes will be lost, ignored, or hidden. In this section we will correct risk measurement by introducing an assumption on how underreporting behaves. In other words, we can say that real operational risk data combine two types of sample selection problems: truncation below a given threshold corresponding to lost data, and random censoring above the level.

Since the reporting process in an organization is central, and underreporting operational risk events is also an operational risk, this should be incorporated in the model. In this chapter we will not discuss the opposite to underreporting, that is, the concept of overreporting. Some may argue that the losses not included in the operational risk calculation will be captured somewhere else in the organization and included in the overall capital figure. However, either if an operational risk loss is reported somewhere else, or not reported at all, we believe that it is essential to adjust for underreporting to obtain a correct operational risk severity distribution for a better risk management.

In Chapter 5 we argued that estimating such an underreporting function from the data itself is a complicated mathematical deconvolution problem and that the rate of convergence of the deconvoluted estimators is often very poor. After extensive interviews with risk expert practitioners and a qualitative decision process, a best guess of an underreporting function has been presented in Chapter 5, but a collection threshold has not been considered there. In this chapter a collection threshold is also included, and therefore, we are only interested in the reporting probabilities above the collection thresholds C_I and C_E.

Let us define two indicator functions $I_I(\cdot)$ and $I_E(\cdot)$ as follows: $I_I(i) = 1$ if X_i^I is reported and zero otherwise, and $I_E(i) = 1$ if X_i^E is reported and zero otherwise. Now, including the minimum collection thresholds, and considering only the truncated samples $X_1^{I,C_I}, X_2^{I,C_I} ..., X_{n_I,C_I}^{I,C_I}$ and $X_1^{E,C_E}, X_2^{E,C_E} ..., X_{n_E,C_E}^{E,C_E}$, we obtain different indicator functions that only evaluate losses reported above these thresholds:

$$
\begin{aligned}
I_{I,C_I}(i) &= 1 \text{ if } X_i^{I,C_I} \text{ is reported and zero otherwise.} \\
I_{E,C_E}(i) &= 1 \text{ if } X_i^{E,C_E} \text{ is reported and zero otherwise.}
\end{aligned}
$$

If the collection thresholds $\{C_I, C_E\}$ are taken into account, it follows that the number of reported losses above the thresholds equals

$$
\begin{aligned}
n_{R,I,C_I} &= \sum_{i=1}^{n_{I,C_I}} I_{I,C_I}(i) \text{ reported number of internal losses above } C_I. \\
n_{R,E,C_E} &= \sum_{i=1}^{n_{E,C_E}} I_{E,C_E}(i) \text{ reported number of external losses above } C_E.
\end{aligned}
$$

Table 6.2: *Quantile values for several models based on simulated data*

Internal Data				
Model	Sample size	$q=0.95$	$q=0.99$	$q=0.995$
$\hat{F}^{\leftarrow}_{R,I}(x_q)$	17	44.59	172.84	289.88
$\hat{F}^{\leftarrow}_{u,I}(x_q)$	17	19.53	48.75	67.50
External Data				
Model	Sample size	$q=0.95$	$q=0.99$	$q=0.995$
$\hat{F}^{\leftarrow}_{R,E}(x_q)$	75	305.64	1271.24	1939.11
$\hat{F}^{\leftarrow}_{u,E}(x_q)$	75	217.78	453.9	594.94

Note: We denote $\hat{F}^{\leftarrow}_{R,d}(x_q)$, with $d=I$ or E, the estimated quantile values for the lognormal models when using only reported data with correction for truncation and no correction for underreporting. $\hat{F}^{\leftarrow}_{u,d}(x_q)$, with $d=I$ or E, denotes estimated quantile values for the lognormal models when using only reported data with correction for truncation and for underreporting.

The number of available data are reduced even further when underreporting is taken into account. Let

$$\left(X_i^{R,I,C_I}\right)_{1\leq i\leq n_{R,I,C_I}}$$

be internal losses over the minimum collection threshold C_I that are reported, and let

$$\left(X_i^{R,E,C_E}\right)_{1\leq i\leq n_{R,E,C_E}}$$

be external losses over the minimum collection threshold C_E that are reported. In our example illustration, $\{n_{I,C_I}, n_{E,C_E}\} = \{17, 75\}$.

Table 6.2 shows what the quantile estimates denoted by $\hat{F}^{\leftarrow}_{R,d}(\cdot)$, with $d=I$ or E, would be when using the underreported sample rather than the truncated one, but no correction is introduced, so that underreporting is ignored.

6.3.1 *Correcting for Underreporting and Collection Threshold*

If we continue to model each data set separately, one should correct each model for underreporting independently. For this we will need to define the underreporting functions by

$$\begin{aligned} u_I(x) &= P(I_{I,C_I}(i)=1|X_i^{I,C_I}=x),\ x\geq C_I, i=1,\ldots,n_{I,C_I} \\ u_E(x) &= P(I_{E,C_E}(i)=1|X_i^{E,C_E}=x),\ x\geq C_E, i=1,\ldots,n_{E,C_E} \end{aligned}$$

Let $f_{O,I}(\cdot)$ and $f_{O,E}(\cdot)$ be the internal and external densities for the occurred losses, respectively, and by using the internal and external densities $f_{R,I}(\cdot)$ and

$f_{R,E}(\cdot)$ for the reported losses, then $f_I(\cdot)$ and $f_E(\cdot)$ are equivalent to

$$f_{u,I}(x) = \frac{f_{R,I}(x)/u_I(x)}{\int_{C_I}^{\infty} f_{R,I}(w)/u_I(w)dw}, \ x \geq C_I, \tag{6.4}$$

and

$$f_{u,E}(x) = \frac{f_{R,E}(x)/u_E(x)}{\int_{C_E}^{\infty} f_{R,E}(w)/u_E(w)dw}, \ x \geq C_E, \tag{6.5}$$

respectively, where the denominator will serve as a normalizing factor. The probability of observing an internal and external operational risk loss above the respective threshold is given by

$$\begin{aligned} P_{I,fo} &= \int_{C_I}^{\infty} f_{O,I}(w)u_I(w)dw \\ P_{E,fo} &= \int_{C_E}^{\infty} f_{O,E}(w)u_E(w)dw. \end{aligned} \tag{6.6}$$

In practice it is quite hard, or sometimes impossible, to establish the external underreporting function. A solution to this would be to build one underreporting function that could be representative of both the internal and external exposure. Being members of a consortium, individuals come together to form this alliance, where members of the consortium are assumed to have similar behaviors. Therefore, assuming that there exists one underreporting function covering both sources intuitively makes sense. A possible underreporting function for the internal and external exposure will then be defined as

$$u(x) = \begin{cases} u_I(x) \text{ if } C_I \leq x < C_E \\ u_E(x) = u_I(x) \text{ if } x \geq C_E, \end{cases}$$

where we have assumed that the external threshold C_E is bigger than the internal threshold C_I.

In order to estimate this function, numerical interpolation and extrapolation is used.

6.3.2 *Including Underreporting in the Estimation Procedure*

Figure 6.2 presents the estimated underreporting function for our example. Here the circles represent experts' opinion on the likelihood of reporting losses of a particular size, and the vertical grey lines represent the thresholds, the dashed line represents the values for the internal underreporting function, while the solid black lines represent the common part. Expert opinion is here defined through the quantiles of the underreporting function.

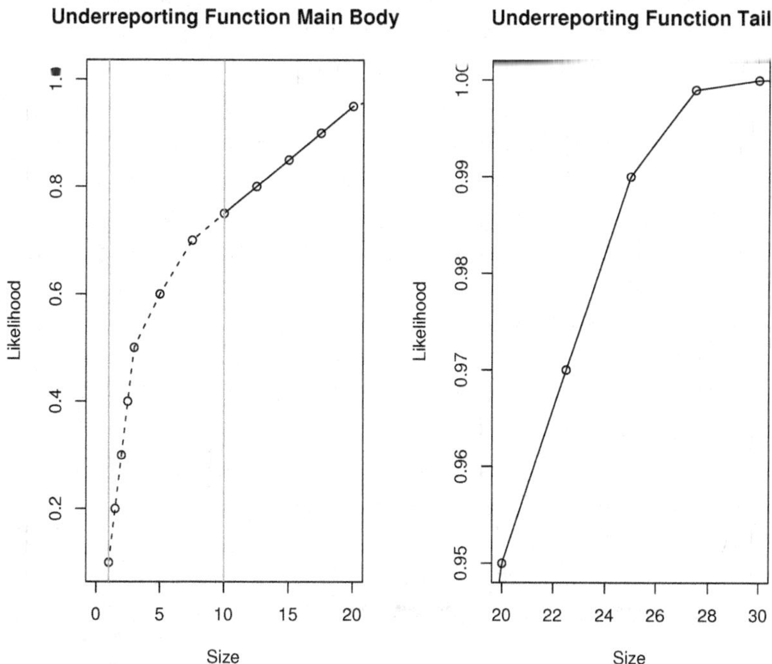

Figure 6.2: *The combined underreporting function.*

With the underreporting function together with $f_{R,I}(\cdot)$ and $f_{R,E}(\cdot)$ available, the original densities $f_{O,I}(\cdot)$ and $f_{O,E}(\cdot)$ can be determined using maximum likelihood estimation, and the results are shown in Figure 6.3.

In Figure 6.3, the three top figures represent the estimated internal densities, and the bottom row the external estimated densities. The domain has been divided in three intervals to facilitate the graphical display. Also presented in Figure 6.3 are the losses and the collection thresholds for each data set. The black solid curves represent the lognormal density (6.1), estimated on the whole samples. The black and grey dashed curves represent the truncated lognormal density (6.3), where the black dashed curves represent the densities estimated on the truncated sample, that is, occurred losses above the thresholds, $\{n_{I,C_I}, n_{E,C_E}\} = \{21, 89\}$. The grey dashed curves are estimated on the underreported samples with total number of reported losses $n_{R,I,C_I}, n_{R,E,C_E}\} = \{17, 75\}$. Here we establish that the internal density becomes

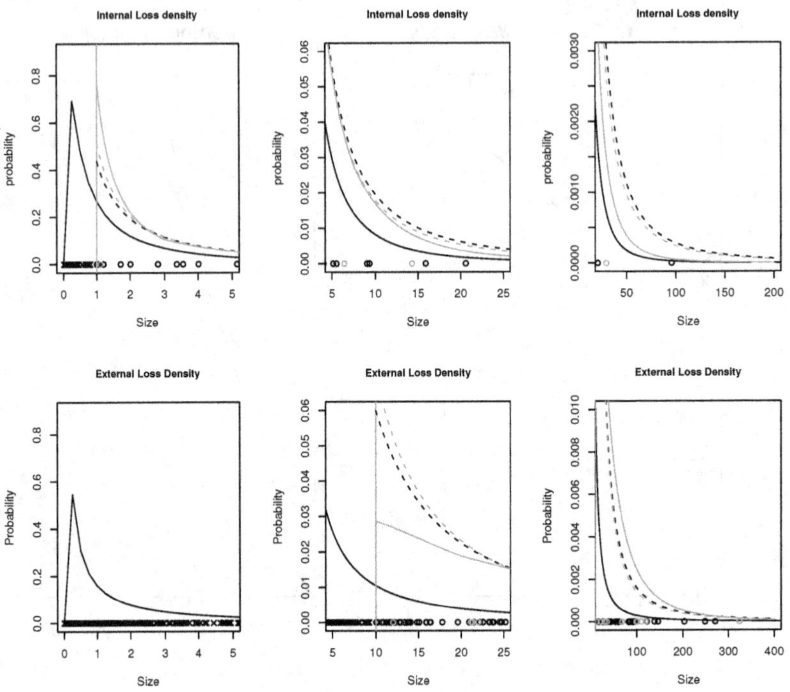

Figure 6.3 *Estimated lognormal densities of the samples* $\left(X_i^I\right)_{1\leq i\leq n_I}$ *and* $\left(X_i^E\right)_{1\leq i\leq n_E}$ *represented by black curves, estimated truncated lognormal densities on the samples* $\left(X_i^{I,C_I}\right)_{1\leq i\leq n_{I,C_I}}$ *and* $\left(X_i^{E,C_E}\right)_{1\leq i\leq n_{E,C_E}}$ *represented by dashed black curves, and estimated truncated lognormal densities on the samples* $\left(X_i^{R,I,C_I}\right)_{1\leq i\leq n_{R,I,C_I}}$ *and* $\left(X_i^{R,E,C_E}\right)_{1\leq i\leq n_{R,E,C_E}}$ *represented by grey dashed curves. The densities represented by grey solid curves are estimated truncated lognormal densities of the samples* $\left(X_i^{R,I,C_I}\right)_{1\leq i\leq n_{R,I,C_I}}$ *and* $\left(X_i^{R,E,C_E}\right)_{1\leq i\leq n_{R,E,C_E}}$*, including underreporting in the estimation.*

lower close to the threshold, while the tail becomes slightly heavier compared to the black dashed curve. This is not that remarkable since four losses are excluded in the estimation procedure, almost all close to the threshold. For the external graphs, the same appearance is found on the grey dashed curve compared to the black dashed curve. However, since there exist more losses, and the underreported losses are spread in size, the differences between the curves are not too large as in the internal situation. The grey solid curves correct for underreporting and are defined by (6.5). Interpreting the outcomes, one could see that adjusting for internal underreporting will provide a lighter tail for this particular simulated sample, while the opposite is found for the external data.

With the estimated densities in (6.5), the probability of observing an operational risk event can be calculated using $P_{I,fo}$ and $P_{E,fo}$. From (6.6) we obtain that $\hat{P}_{I,fo} \cong 44\%$ and $\hat{P}_{E,fo} \cong 96\%$. These estimated values seem reasonable if we take the mean of the reported expert opinions and we obtain 68% on the support $[C_I, \infty)$ and 93% on the support $[C_E, \infty)$. Of course, the differences 44% to 63% and 96% to 93% depends on the data availability when estimating the densities in (6.5). The upper quantiles are presented in Table 6.2.

The second row of Table 6.2 for the internal and the external data, respectively, presents the models correcting for truncation and underreporting, and it is represented by $\hat{\overset{\leftarrow}{\hat{F}}}_{O,d}(\cdot)$, with $d = I$ or E. These results, when compared to the first corresponding row for internal and external data shown in Table 6.1, provide similar values. As a consequence, this shows that the correction for truncation and for underreporting provides results that are close to the original severity distribution of the original data.

6.4 A Mixing Model in a Truncation Framework

From this section onwards, we will combine internal and external data. Again we must emphasize that distributional assumptions can be modified in practice, and the ones assumed here are not necessarily the best choices for every applied situation.

A company should combine internal and external data for internal loss distribution modeling, when the sample of internal data seems insufficient to represent all the domain accurately. Here we take into account both minimum collection threshold as well as underreporting. The mixing model should utilize the external data as prior knowledge. Then, a nonparametric smoothing technique adjustment should be applied on the prior knowledge according to internal data availability.

In our example, the mixing model is applied as a first step. Let us assume all occurred internal losses $X_1^I, X_2^I.., X_{n_I}^I$ were available. Then they could be transformed to bounded support by

$$\left(\hat{F}_{O,E}(X_i^I)\right)_{1 \le i \le n_I} \in [0,1],$$

where $\hat{F}_{O,E}(\cdot)$ denotes a lognormal cumulative distribution function, and its parameters have been estimated with the sample of occurred external losses (assuming all of them were also available). The mixing methodology continues by bring into play kernel density estimation on the bounded support, and thereby a correction function will be estimated that depends on the internal data characteristics and its discrepancy versus prior knowledge. The underlying parametric transformation is guided by a local constant kernel density estimator presented by

$$\hat{h}(\hat{F}_{O,E}(x)) = \frac{1}{n_I \cdot \alpha_{01}(\hat{F}_{O,E}(x), b)} \sum_{i=1}^{n_I} K_b\left(\hat{F}_{O,E}(X_i^I) - \hat{F}_{O,E}(x)\right),$$

where $\hat{F}_{O,E}(x) \in [0,1]$ and losses are transformed, that is, $\hat{F}_{O,E}(X_i^I)$. We assume a symmetric Epanechnikov kernel K_b with $K_b(\cdot) = (1/b)K(\cdot/b)$, Silverman's rule-of-thumb bandwidth b is estimated on the internal sample, and the boundary correction $\alpha_{ij}(\cdot,\cdot)$ is defined in Chapter 3.

The final step in the transformation process is to backtransform to the original axis. This is done by transforming the smoothed distribution $\hat{k}(\cdot)$ of the transformed data with $\hat{F}_{O,E}^{\leftarrow}(\cdot)$. The transformation methodology can be summarized in one explicit expression:

$$\begin{aligned} \hat{f}_m(x) &= \frac{\hat{f}_{O,E}(x)}{n_I \cdot \alpha_{01}(\hat{F}_{O,E}(x), b)} \sum_{i=1}^{n_I} K_b\left(\hat{F}_{O,E}(X_i^I) - \hat{F}_{O,E}(x)\right) & (6.7) \\ &= \hat{f}_{O,E}(x) \cdot \hat{h}(\hat{F}_{O,E}(x)),\ x \in [0,\infty), & (6.8) \end{aligned}$$

where subindex m indicates a mixing model. The transformation methodology has been presented in Chapter 3.

When incorporating the thresholds on occurred data, that is, utilizing the data $X_1^{I,C_I}, X_2^{I,C_I}, \ldots, X_{n_{I,C_I}}^{I,C_I}$ and $X_1^{E,C_E}, X_2^{E,C_E}, \ldots, X_{n_{E,C_E}}^{E,C_E}$, some adjustment is needed in the transformation process. In practice, as we should see in the application section if C_I and C_E are equal and there would not be a transformation problem. However, in our simulated example $C_I \neq C_E$, the transformation function will be defined on the support $[C_E,\infty)$, while the occurred internal data that should be incorporated in the transformation function are defined on the support $[C_I,\infty)$. Therefore, the losses in the support $[C_I, C_E]$ cannot be transformed correctly since the transformation function is not defined on this support. The procedure to solve this issue draws on extrapolation from the lower limit (C_I) to the upper threshold (boundary point C_E).

Let us assume that $F_E(\cdot)$ is estimated using external data; then we will extrapolate and normalize to the interval $[C_I,\infty)$ to solve the transformation problem. By defining the extrapolated normalized external density as

$$\hat{f}_E^*(x) = \frac{\hat{f}_E(x)}{\int_{C_I}^{\infty} \hat{f}_E(w)dw},\ x \geq C_I, \tag{6.9}$$

we could determine a truncated cumulative distribution function $\hat{F}_E^*(\cdot)$ that should transform the internal occurred losses $X_1^{I,C_I}, X_2^{I,C_I}, \ldots, X_{n_{I,C_I}}^{I,C_I}$ to bounded support. That is,

$$\left(\hat{F}_E^*(X_i^{I,C_I})\right)_{1\leq i\leq n_{I,C_I}} \in [\hat{F}_E^*(C_I), 1],$$

Table 6.3: *Quantile values for several models based on pooled simulated data*

		Probability q		
Model	Sample size	0.95	0.99	0.995
$\hat{F}_m^{\leftarrow}(x_q)$	600	36.26	213.46	410.50
$\hat{F}_m^{\leftarrow *}(x_q)$	110	115.54	485.77	830.36
$\hat{F}_{R,m}^{\leftarrow *}(x_q)$	92	96.27	472.95	873.05
$\hat{F}_{O,m}^{\leftarrow *}(x_q)$	92	65.28	279.43	304.75

where $\hat{F}_E^*(C_I)$ represent the estimated internal threshold on the transformed scale. The truncated mixing semiparametric model will then follow the expression

$$\begin{aligned} \hat{f}_m^*(x) &= \frac{\hat{f}_E^*(x)}{n_{I,C_I} \cdot \alpha_{01}(\hat{F}_E^*(x),b)} \sum_{i=1}^{n_{I,C_I}} K_b\left(\hat{F}_E^*(X_i^{I,C_I}) - \hat{F}_E^*(x)\right) \\ &= \hat{f}_E^*(x) \cdot \hat{h}(\hat{F}_E^*(x)),\ x \in [C_I, \infty). \end{aligned} \tag{6.10}$$

If collection thresholds were identical, $C_I = C_E$, the distribution $\hat{F}_E^*(\cdot)$ could be replaced with $\hat{F}_E(\cdot)$ in the above expression above. The final step is to correct for underreporting. Following the same argumentation as above, we evaluate the data set $X_1^{R,I,C_I}, X_2^{R,I,C_I}, \ldots, X_{n_{R,I,C_I}}^{R,I,C_I}$ instead of $X_1^{I,C_I}, X_2^{I,C_I}, \ldots, X_{n_{I,C_I}}^{I,C_I}$ in the bounded support, guided by the transformation function $\hat{F}_{R,E}^*(\cdot)$, obtained by following a procedure that is similar to (6.9).

Hence, the final model becomes

$$\begin{aligned} \hat{f}_{R,m}^*(x) &= \frac{\hat{f}_{R,E}^*(x)}{n_{R,I,C_I} \cdot \alpha_{01}(\hat{F}_{R,E}^*(x),b)} \sum_{i=1}^{n_{R,I,C_I}} K_b\left(\hat{F}_{R,E}^*(X_i^{R,I,C_I}) - \hat{F}_{R,E}^*(x)\right) \\ &= \hat{f}_{R,E}^*(x) \cdot \hat{h}(\hat{F}_{R,E}^*(x)),\ x \in [C_I, \infty), \end{aligned} \tag{6.11}$$

where the kernel density estimation function $\hat{k}$ is defined on $[\hat{F}_{R,E}^*(C_I), 1]$. In Table 6.3, the three different mixing models (4.3), for the whole sample of occurred losses, (6.10) with the truncated sample and (6.11) with the truncated and underreported sample are compared in the first three rows for different quantile values, denoted by $\hat{F}_m^{\leftarrow}(\cdot)$, $\hat{F}_m^{\leftarrow *}(\cdot)$, and $\hat{F}_{R,m}^{\leftarrow *}(\cdot)$.

Until this point in the chapter, the two methodologies, namely, correction for underreporting and mixing internal and external data, have been considered separately. In the remaining part of this section we determine a model that incorporates both these adjustments. The procedure begins by seeking the transformation function. This transformation function should correct for underreporting as well as using the internal threshold C_I as lower bound. We estimate a normalized density extrapolated for the extended support $[C_I, \infty)$ as

$$\hat{f}^*_{O,E}(x) = \frac{\hat{f}_{O,E}(x)}{\int_{C_I}^{\infty} \hat{f}_{O,E}(w)dw}, \; x \in [c_I, \infty). \tag{6.12}$$

Using $\hat{F}^*_{O,E}(\cdot)$, the cumulative distribution function derived from (6.12), we transform the losses $X_1^{R,I,C_I}, X_2^{R,I,C_I}, \ldots, X_{n_{R,I,C_I}}^{R,I,C_I}$ to bounded support. The final semiparametric estimator is then

$$\begin{aligned}
\hat{f}^*_{O,m}(x) &= \\
&= \frac{\hat{f}^*_{O,E}(x)}{n_{R,I,C_I} \cdot \alpha_{01}(\hat{F}^*_{O,E}(x), b)} \sum_{i=1}^{n_{R,I,C_I}} K_b\left(\hat{F}^*_{O,E}(X_i^{R,I,C_I}) - \hat{F}^*_{O,E}(x)\right) \\
&= \hat{f}^*_{O,E}(x) \cdot \hat{h}(\hat{F}^*_{O,E}(x)), \; x \in [C_I, \infty),
\end{aligned} \tag{6.13}$$

where the kernel density estimation function $\hat{k}(\cdot)$ is defined on $[\hat{F}^*_{O,E}(C_I), 1]$. The quantiles following from model (6.13) with the underreported sample are also presented in the last row of Table 6.3.

It is interesting to see is the outcome of model (6.13) in the tail. Focusing on the quantile 99.5%, the proposed model $\hat{F}^{\leftarrow *}_{O,m}(x_q)$ shows the smallest value among the four mixing models. However, this is not the case for the lower quantiles 95% and 99%. The proposed model presents larger values on all quantiles in comparison to the pure internal models and smaller values, as compared to the pure external models.

Figure 6.4 plots the simulated severity distributions with the models that can be used when pooling together internal and external data. The grey and dashed grey line correspond to modeling internal and external data separately. They are plotted as a reference.

6.5 Operational Risk Application

This section shows how the choice of methodology can significantly affect the estimation of required risk-based capital. We perform a LDA (Loss Distribution Analysis) on real operational losses and calculate the risk measures Value-at-Risk (VaR) and Tail-Value-at-Risk (TVaR) for different levels of risk tolerance α. The operational loss data follows the event risk category *Internal Fraud*, and the internal losses are defined by $X_1^{R,I,C_I}, X_2^{R,I,C_I}, \ldots, X_{n_{R,I,C_I}}^{R,I,C_I}$, and the external data $X_1^{R,E,C_E}, X_2^{R,E,C_E}, \ldots, X_{n_{R,E,C_E}}^{R,E,C_E}$ are taken from a consortium. The collection threshold for the two samples are the same with the value

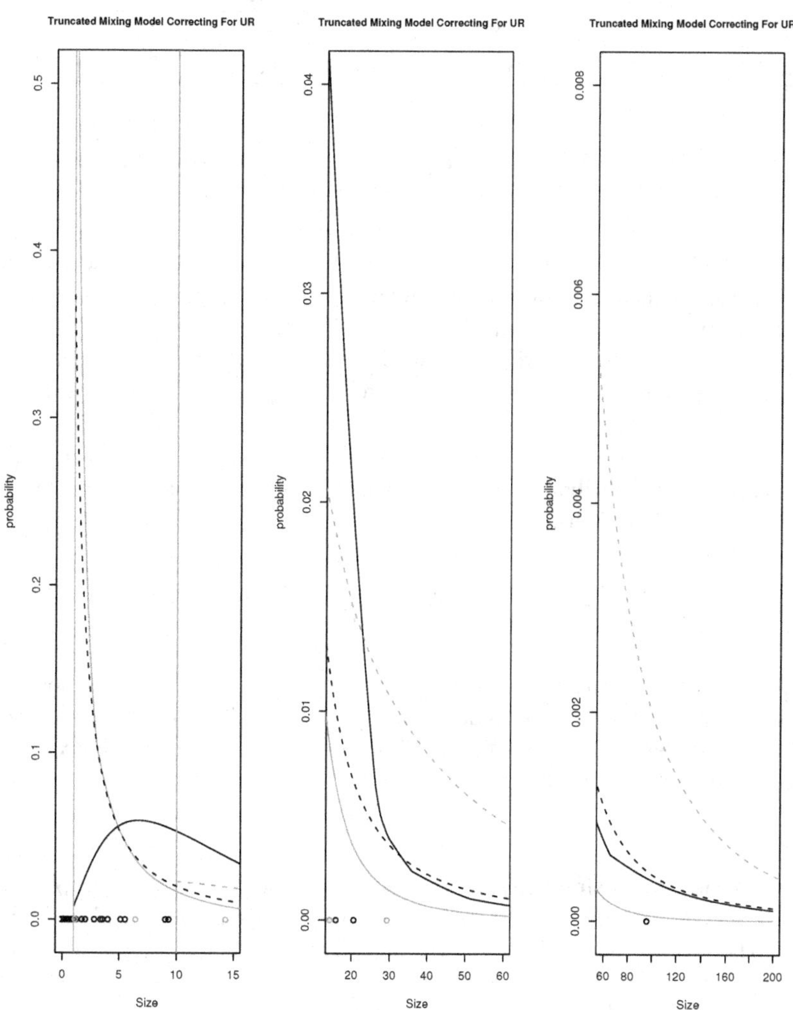

Figure 6.4 *Model (6.13) is represented by a black curve, model (6.10) is represented by a dashed black curve, and internal and external versions of model (6.5) are represented by a grey and dashed grey curves, respectively.*

Table 6.4: *Statistics for Event Risk Category* Internal Fraud

	Reported Losses	Maximum Loss	Sample Mean	Sample Median	Standard Deviation	Years
X_i^{R,I,C_I}	111	1.993	0.105	0.033	0.239	2
X_i^{R,E,C_E}	227	36.36	0.957	0.128	3.201	7

$C_I = C_E = C = 10.000$£. A summary of the operational losses is presented in Table 6.4.

In total, eight models will be compared and evaluated. The developed model (6.13), the truncated mixing model (6.10), internal and external underreporting models (6.5), truncated internal and truncated external models such as (6.3), and pure parametric model without any adjustments. The underlying distribution is important both in the underreporting procedure as well as in the mixing methodology.

The prior density that will be used in this study is the generalized Champernowne distribution (GCD), which has been introduced in Chapter 2. Its density takes the form

$$f_{\alpha,M,c}(x) = \frac{\alpha\,(x+c)^{\alpha-1}\left((M+c)^{\alpha}-c^{\alpha}\right)}{\left((x+c)^{\alpha}+(M+c)^{\alpha}-2c^{\alpha}\right)^{2}}, \ \forall x \in \mathbb{R}_{+}, \alpha > 0, M > 0, c \geq 0. \tag{6.14}$$

This density will then be estimated on both samples, used separately as models, and incorporated both in the underreporting models and the mixing methodology. The underreporting function used in this study is estimated by experts' opinion and described by Figure 6.5.

For the sake of simplicity, we introduce the abbreviations $\hat{F}_1, \hat{F}_2, ..., \hat{F}_8$ for the estimated severity models considered in the analysis, described in Table 6.5. The severity models $\hat{F}_1$ and $\hat{F}_2$ are pure parametric models estimated on the internal and external data, respectively, using the parametric choice (6.14). The two models $\hat{F}_3$ and $\hat{F}_4$ are pure parametric as well; however, they are defined by a truncated version of (6.14) outlined by formula (6.3). Further, the models $\hat{F}_5$ and $\hat{F}_6$ are offsprings from the two previous models but with the differences that the underreporting function defined by Figure 6.5 is taken into account by using formula (6.5). Model $\hat{F}_7$ is the first mixing model considered in the study with the prior start (6.14) and defined through equation (6.10). The final model, $\hat{F}_8$, is an extended version of $\hat{F}_7$ by also including underreporting in the estimation.

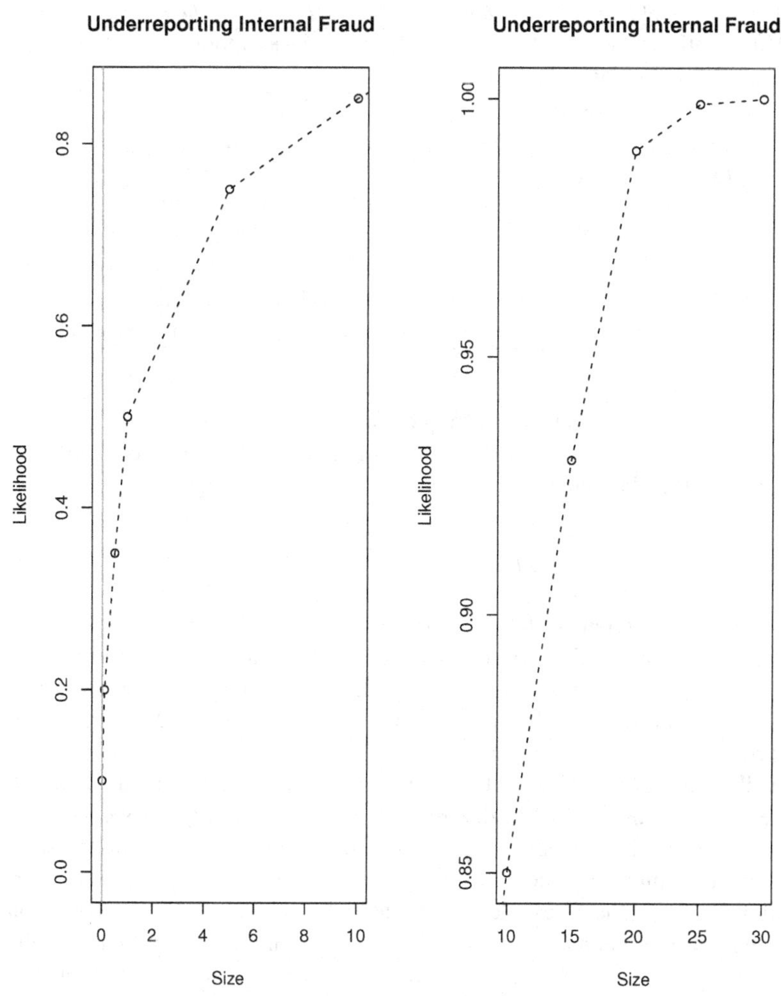

Figure 6.5: *Underreporting function for the event risk category* Internal fraud.

Three different estimation procedures are needed as a first step to estimate the eight models. The parameter M in the distribution (6.14) is estimated by the empirical median in all three situations. Hence, $F_{\alpha,M,c}(M) = 1/2$, thereby indicating that the empirical median is an appropriate estimation choice. The other parameters $\{\alpha, c\}$ are then obtained by maximizing the log-likelihood function corresponding to each model.

Table 6.5: *Abbreviations of details on the severity choices*

Abbreviation	Internal data	External data	Pooled data	Description
$\hat{F}_1$	(6.14)	·	·	Pure parametric
$\hat{F}_2$	·	(6.14)	·	Pure parametric
$\hat{F}_3$	(6.3)	·	·	Trunc. parametric
$\hat{F}_4$	·	(6.3)	·	Trunc. parametric
$\hat{F}_5$	(6.5)	·	·	Trunc. und.rep. parametric
$\hat{F}_6$	·	(6.5)	·	Trunc. und.rep. parametric
$\hat{F}_7$	·	·	(6.10)	Trunc. mixing model
$\hat{F}_8$	·	·	(6.13)	Trunc. und.rep. mixing model

We are going to simulate the frequency of events. In the Monte Carlo simulation, N_{R,I,C_I} is assumed to follow a homogenous Poisson process with intensity $\lambda > 0$ and defined by

$$P\left(N_{R,I,C_I} = n\right) = e^{-\lambda t}\frac{(\lambda t)^n}{n!}.$$

We estimate the intensity for each model and use maximum likelihood. So, $\hat{\lambda} = n/T$, where n is the reported number of losses and T is the corresponding collection period presented in Table 6.4. This frequency estimation is then incorporated in five models, the ones without underreporting. Model 5, 6 and 8 correct for underreporting and hence, the frequency needs to be corrected as well as the severity side. With underreporting, we assume that the reported losses N_{I,C_I} follow a Poisson distribution with intensity equal to λP, where P is the probability of observing an operational risk loss, defined by (6.6). When accounting for underreporting, we focus on N_{I,C_I}, the occurred internal losses above the threshold C_I. We assume that N_{I,C_I} follows a Poisson distribution with intensity λ. The correction is then obtained for each of the three underreporting models by dividing the estimated $\hat{\lambda}$ with the probability of observing a loss for each model. When we calculate these probabilities of observing an operational risk loss using (6.6), we get $\{\hat{P}_5, \hat{P}_6, \hat{P}_8\} = \{15.1\%, 28.6\%, 18.0\%\}$. Then, all the intensities used in this study will be $\{\hat{\lambda}_5, \hat{\lambda}_6, \hat{\lambda}_8\} = \{366.7, 193.8, 308.3\}$.

The aggregated total loss S_k is then defined as the stochastic process

$$S_k = \sum_{i=1}^{N_{R,I,C_I}} \widehat{F}_k^{\leftarrow}(u_i)$$

for severity model assumption k. Here, u_i is sampled from a uniform distribution guided by sampling the Poisson with intensity equal to one of the four estimated models. Function $\widehat{F}_k^{\leftarrow}(\cdot)$ is the estimated inverse function

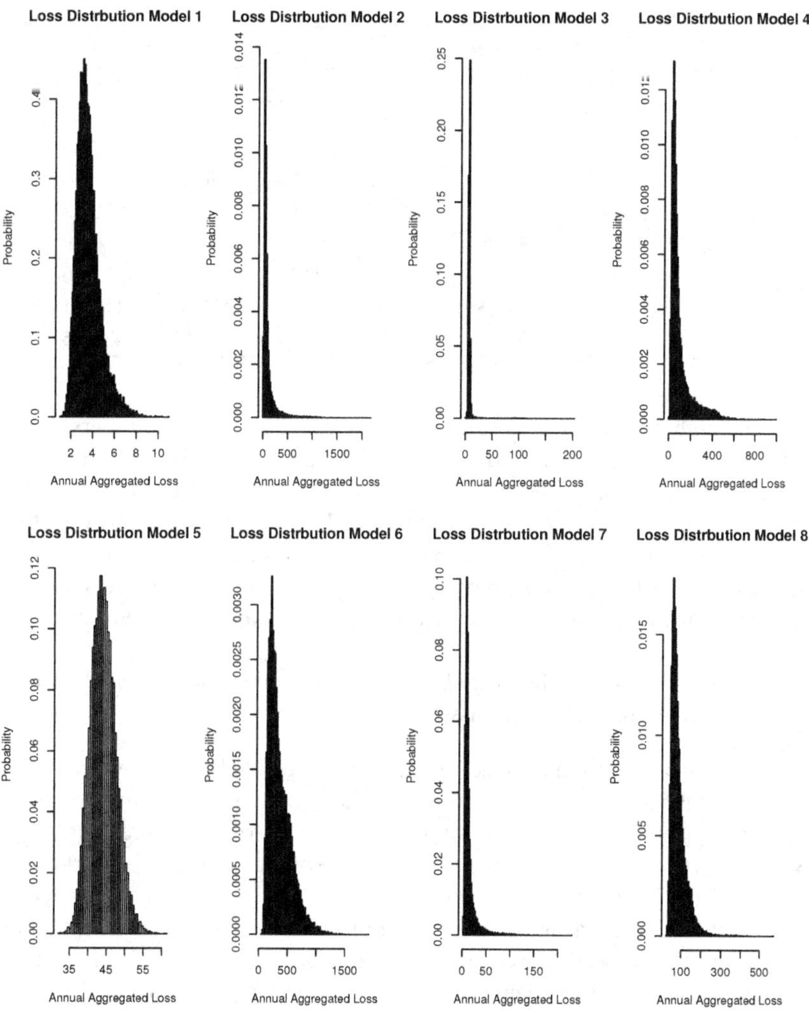

Figure 6.6 *The estimated total loss distributions* S_k, $k = 1, \ldots, 8$, *for the event risk category* Internal fraud.

for one of the eight models described in Table 6.5. This procedure is then repeated 20,000 times for each model, resulting in an aggregated empirical loss distribution with holding period one year. This is illustrated visually in Figure 6.6.

Table 6.6: *Risk measures for the different loss distributions*

Model	Value-at-Risk			Tail-Value-at-Risk		
	95%	99%	99.9%	95%	99%	99.9%
S_1	5.78	7.20	8.68	6.64	7.94	9.86
S_2	388.41	880.81	1177.80	664.37	1052.54	1427.06
S_3	9.75	32.04	109.60	25.88	76.24	122.66
S_4	345.89	487.61	682.33	439.08	578.63	812.48
S_5	50.13	53.08	56.51	51.98	54.88	60.69
S_6	749.75	1004.13	1276.37	902.74	1134.83	1483.66
S_7	44.01	91.25	128.46	71.04	109.53	152.25
S_8	160.27	229.45	377.99	206.47	299.73	438.65

The eighth model seems to place its outcome between the others. Interpreting the outcomes, we find that S_1 and S_5 provide the lightest tails. It is interesting to see that S_1 becomes more heavy tailed when truncation is incorporated (see S_3), and lighter when adding underreporting. The same effect, in some sense, holds when we study S_2, S_4, and S_6. The effect of adding underreporting to S_7 leads to obtaining the proposed S_8. However, if we study the more extreme outcomes with the risk measures VaR and TVaR, we find some effects.

In Table 6.6, some chosen risk levels are specified for the risk measures VaR and TVaR for each model assumption.

The data application provided us with insight on the behavior when adding more sophisticated corrections to the models. Starting from two parametric models S_1 and S_2, we see that the outcome on the risk measure $\text{VaR}_{99.9\%}$ was over 100 times larger value using the external data source instead of the internal data. By including truncation in S_3 and S_4, the internal model provided a heavier tail, while a lighter was found on the external data model. Adding underreporting as well in S_5 placed the result between models for S_1 and S_3, while S_6 became larger than both S_2 and S_4. By using the truncated mixing model in S_7, the prior knowledge of the estimation brings a larger value than the internal data models S_1, S_3, and S_5. However, we could see that the outcome is far away from the external data models. When taking underreporting into account, the outcome from the model in S_8 increases by 294% compared to S_7.

6.6 Further Reading and Bibliographic Notes

This chapter presented several systematic correction techniques essential to quantify a company's operational risk exposure. Starting with internal and external data sets, the importance of including collection thresholds was shown. Thereafter, discussion and illustration on the difference between modeling observed or reported losses was given. The development continued to combine two data sources with a final result that both adjusted for the collection threshold and underreporting.

Reference [52] has much more information on continuous and particularly truncated probability density functions.

The grounds for loss distribution modeling can be found in references [56], [60], [18], and [62]. The past few years have provided the literature with alternative models on how one can incorporate external information into internal loss distribution modeling. The papers by Sheuchenko and Wthrich [69], Bhlmann et al. [12], and Lambrigger et al. [59] make use of Bayesian inference to combine several sources of data. Dahen and Dionne [23] present methods to scale severities and frequencies in the context of banks because evidence has been found of enormous heterogeneity due to diversity in size of the firm and location operations. Figini et al. [35] develop a method to mix data by applying a truncated external data for internal loss distribution modeling. Another procedure was proposed in reference [75], where Bayesian credibility is used to combine external and internal data. Gustafsson and Nielsen [47], and the extended version by Gustaffson [45], develop a systematic approach that incorporates external information into internal loss distribution modeling. The standard statistical model resembles Bayesian methodology and credibility theory in the sense that prior knowledge (external data) has more weight when internal data are scarce than when internal data are abundant.

A simplified approach was introduced by Gullén et al. [43] and an extended version by Buch-Kromann et al. [7]. They developed an underreporting function in a way that the theoretical problem is simplified, and they proposed a solution that is closely related to what one would have obtained if all internal losses would have been observed without underreporting.

For a review of modern kernel smoothing techniques, see the fundamental text by Wand and Jones [73], and for regression functions, see reference [34].

In recent years, new and more advanced parametric distributions for operational risk have emerged, with more parameters and more advanced methods of estimation. These more advanced parametric distributions present a very reasonable goodness of fit in practical situations. However, using parametric estimators requires that practitioners need to reconsider the parametric family when updating data. Some new distributions are the α-stable distribution and the g-and-h distribution. The α-stable distribution is investigated in reference [41] where they test the distribution by transforming the data to better fit the parameter estimation. They show that, in order to get a good fit of the operational risk data, they are forced to model the body and the tail separately when using the α-stable distribution. We find this troubling since we are interested in a parametric distribution that fit the entire data. There is another problem with the α-stable distribution, and it is that closed-form densities do not exist for most cases. Hence, parameter estimations based on the maximum likelihood method becomes complicated, a strong reason for practitioners not to choose this distribution. The g-and-h distribution, on the other hand, has an appealing link to the common extreme value theory (EVT) methodology, which is addressed in detail by Degen et al.[27]. However, the *g*-and-*h* distribution has a flaw in its slow convergence to the generalized Pareto distribution. Using it in connection with EVT may lead to inaccurate capital estimations, in particular for small levels of risk tolerance, which is shown in reference [28]. Instead, we used the GCD. The GCD has the same appealing property as the *g*-and-*h* distribution with its link to EVT. However, the tail's convergence to the heavy-tailed Pareto distribution is faster than the *g*-and-*h* distribution, which favors the GCD. Also, a large flexibility is imbued in the distribution (see, for example, reference [49]).

Chapter 7

A Guided Practical Example

7.1 Introduction

A series of data sets have been created to help practitioners in the implementation of the methods described throughout the book. The basic procedures have been programmed in SAS© and in R, and the results are shown in order to discuss practical application and the interpretation of results. The data sets can be obtained from the authors upon request.

Table 7.1 presents a list of the data that are being used in this chapter. The first two data sets correspond to operational risk data from internal and external sources. These data have been generated for educational purposes, and they do not refer to any particular company or type of operational risk. We will call these data the *internal data set* and the *external data set*, respectively. Further, for illustrative purposes, we also present two other sets of data that are similar to operational risk data that can be found in the publicly available operational risk sources. These are called *public data risk no. 1* and *public data risk no. 2*, respectively.

7.2 Descriptive Statistics and Basic Procedures in SAS©

Data are stored in Microsoft Excel© or ASCII text files and can easily be imported into SAS© data sets.

The following program assumes that the *internal data set* has been stored in library `oper` and is called `Int_sim_table2_1`. Loss amounts are stored in variable y. The program calculates basic statistics, plots a histogram, and overlays the estimated parametric densities. The parametric models presented in

Table 7.1: *Data sets used in this chapter*

Name	Content of operational risk loss data
Internal data set	75 observed loss amounts
External data set	700 observed loss amounts
Public data risk no. 1	1000 observed loss amounts for category no. 1
Public data risk no. 2	400 observed loss amounts for category no. 2

Note: These data sets are not the ones used in previous chapters.

Chapter 2 are implemented. Parameter estimates and goodness-of-fit measures are also obtained.

```
proc univariate data=oper.Int_sim_table2_1;
  var y;
  histogram / exponential
              lognormal
              weibull
              gamma;
run;
```

In order to obtain basic descriptive statistics for the *external data set*, one needs to change `Int_sim_table2_1` by `Ext_sim_table2_1` so that the *external data set* is used instead of the *internal data set*. Therefore, only the data set name that is going to be used needs to be changed in the program. In the rest of this chapter, our example program are not going to be repeated for each data set as it is trivial to modify them for every different set of observations.

The results using the program for descriptive statistics are listed in Table 7.2 for the *internal* and *external data set*. More precisely, Table 7.2 and Table 7.3 provide basic descriptive statistics. Table 7.4 presents location tests, and Table 7.5 has 11 quantile values for each data set. Table 7.6 highlights extreme events, both the lowest and the highest for each data set.

In Chapter 2 we introduced several parametric distributions for modeling operational risk. We give some results for these distributions for the *internal* and *external data set*. In Tables 7.7 to 7.17, the estimated parameters, goodness-of-fit tests, and estimated quantile values are provided for the lognormal, exponential, Weibull, and Gamma distributions, which were presented in Chapter 2.

All the goodness-of-fit tests reject, with the usual confidence level at 95%, the parametric distributional assumption given by the parametric shapes that have been implemented in the first program. This means that the shape of the true density that has generated the data does not conform to those popular parametric models. This is the case both for the *internal* and *external* data set.

In order to fit the logistic distribution, the SAS© program code is given. This program implements the method of moments estimation procedure.

Table 7.2: *The Univariate procedure: Output for guided example data*

	Internal data	External data
N	75	700
Mean	0.1756	0.6788
Std Deviation	0.2777	4.0937
Skewness	3.7701	9.7909
Uncorrected SS	8.0163	12036.7155
Coeff Variation	158.1518	603.0403
Sum Weights	75	700
Sum Observations	13.167	475.191
Variance	0.0771	16.7584
Kurtosis	16.5902	108.3463
Corrected SS	5.7047	11714.1348
Std Error Mean	0.0321	0.1547

```
/*PARAMETRIC ESTIMATION Logistic*/
proc iml;
  /*Read Internal and External data*/
  use oper.int_sim_table2_1 var{y};
  read all into x_int;
  use oper.ext_sim_table2_1 var{y};
  read all into x_ext;
  /*Modify the following line to select the
    Internal or the External data set*/
  x=x_int;
  /* nr is the number of observations in the sample*/
  nr=nrow(x);
  /*Calculate method of moments estimation*/
  par=j(2,1,0);
  par[1]=sum(x)/nr;
  s2=sum((x-par[1])##2)/nr;
  par[2]=sqrt(3*s2)*(1/3.14159265359);
  Label={'mu', 'sigma'};
  print Label par;
quit;
```

The logistic parameter estimates for the *internal data set* and for the *external data set* are listed in Table 7.18.

Table 7.3: *Basic statistical measures for guided example data*

Location	Internal data	External data
Mean	0.175560	0.678844
Median	0.081000	0.033000
Mode	0.069000	0.013000
Variability	**Internal data**	**External data**
Std Deviation	0.27765	4.09370
Variance	0.07709	16.75842
Range	1.77000	52.12900
Interquartile Range	0.15200	0.04700

Table 7.4: *Tests for location:* $\mu_0 = 0$ *for guided example data*

	Internal data	
Test	Statistic	p Value
Student's t (t)	5.475913	$Pr > \|t\|$ <.0001
Sign (M)	37.5	$Pr \geq \|M\|$ <.0001
Signed Rank (S)	1425	$Pr \geq \|S\|$ <.0001
	External data	
Test	Statistic	p Value
Student's t (t)	4.387354	$Pr > \|t\|$ <.0001
Sign (M)	350	$Pr \geq \|M\|$ <.0001
Signed Rank (S)	122675	$Pr \geq \|S\|$ <.0001

Table 7.5: *Quantiles for the guided example data*

Quantile	Internal data	External data
100% Max	1.773	52.130
99%	1.773	19.439
95%	0.780	1.672
90%	0.390	0.249
75% Q3	0.201	0.063
50% Median	0.081	0.033
25% Q1	0.049	0.016
10%	0.030	0.010
5%	0.021	0.007
1%	0.003	0.003
0% Min	0.003	0.001

Table 7.6: *Extreme observations for the guided example data*

Internal data			
Lowest		Highest	
Value	Observation	Value	Observation
0.003	2	0.730	35
0.013	28	0.780	9
0.016	45	1.056	59
0.021	73	1.068	3
0.026	37	1.773	18
External data			
Lowest		Highest	
Value	Observation	Value	Observation
0.001	661	28.722	275
0.001	301	28.729	366
0.002	424	49.812	487
0.002	262	52.129	150
0.002	190	52.130	427

Table 7.7: *Fitted lognormal distribution for the guided example data*

Internal data		
Parameter	Symbol	Estimate
Threshold	Theta	0
Scale	Zeta	−2.36353
Shape	Sigma	1.06294
Mean		0.165529
Std Dev		0.239598
External data		
Parameter	Symbol	Estimate
Threshold	Theta	0
Scale	Zeta	−3.17383
Shape	Sigma	1.626237
Mean		0.157001
Std Dev		0.567782

Note: SAS parametrization.

Table 7.8 *Goodness-of-fit tests for the lognormal distribution in the guided example data*

	Internal data			
Test	Statistic		p Value	
Kolmogorov–Smirnov	D	0.13178203	$Pr > D$	<0.010
Cramer–von Mises	W-Sq	0.16893055	$Pr > W - Sq$	0.014
Anderson–Darling	A-Sq	0.96915484	$Pr > A - Sq$	0.015
	External data			
Test	Statistic		p Value	
Kolmogorov–Smirnov	D	0.1566787	$Pr > D$	<0.010
Cramer–von Mises	W-Sq	4.8968920	$Pr > W - Sq$	<0.005
Anderson–Darling	A-Sq	29.2290150	$Pr > A - Sq$	<0.005

Table 7.9: *Quantiles for lognormal distribution in the guided example data*

Percent	Internal data		External data	
	Observed	Estimated	Observed	Estimated
1.0	0.003	0.00794	0.003	0.00095
5.0	0.021	0.01638	0.007	0.00288
10.0	0.030	0.02410	0.010	0.00521
25.0	0.049	0.04594	0.016	0.01397
50.0	0.081	0.09409	0.033	0.04184
75.0	0.201	0.19271	0.063	0.12531
90.0	0.390	0.36739	0.249	0.33631
95.0	0.780	0.54056	1.672	0.60720
99.0	1.773	1.11543	19.439	1.83927

Table 7.10: *Fitted exponential distribution for the guided example data*

Internal data		
Parameter	Symbol	Estimate
Threshold	Theta	0
Scale	Sigma	0.17556
Mean		0.17556
Std Dev		0.17556
External data		
Parameter	Symbol	Estimate
Threshold	Theta	0
Scale	Sigma	0.678844
Mean		0.678844
Std Dev		0.678844

Note: SAS parametrization.

Table 7.11 *Goodness-of-fit tests for the exponential distribution in the guided example data*

		Internal data		
Test	Statistic		p Value	
Kolmogorov–Smirnov	D	0.22608447	$Pr > D$	<0.010
Cramer–von Mises	W-Sq	0.70510305	$Pr > W - Sq$	<0.010
Anderson–Darling	A-Sq	3.66848810	$Pr > A - Sq$	<0.010
		External data		
Test	Statistic		p Value	
Kolmogorov–Smirnov	D	0.704456	$Pr > D$	<0.010
Cramer–von Mises	W-Sq	133.803470	$Pr > W - Sq$	<0.010
Anderson–Darling	A-Sq	946.734147	$Pr > A - Sq$	<0.010

Table 7.12: *Quantiles for exponential distribution in the guided example data*

Percent	Internal data		External data	
	Observed	Estimated	Observed	Estimated
1.0	0.003	0.00176	0.003	0.00682
5.0	0.021	0.00901	0.007	0.03482
10.0	0.030	0.01850	0.010	0.07152
25.0	0.049	0.05051	0.016	0.19529
50.0	0.081	0.12169	0.033	0.47054
75.0	0.201	0.24338	0.063	0.94108
90.0	0.390	0.40424	0.249	1.56310
95.0	0.780	0.52593	1.672	2.03364
99.0	1.773	0.80848	19.439	3.12619

Table 7.13: *Fitted Weibull distribution for the guided example data*

	Internal data	
Parameter	Symbol	Estimate
Threshold	Theta	0
Scale	Sigma	0.161968
Shape	C	0.88066
Mean		0.172509
Std Dev		0.196385
	External data	
Parameter	Symbol	Estimate
Threshold	Theta	0
Scale	Sigma	0.105862
Shape	C	0.441553
Mean		0.273956
Std Dev		0.73493

Note: SAS parametrization.

Table 7.14 *Goodness-of-fit tests for the Weibull distribution in the guided example data*

	Internal Data			
Test	Statistic		p Value	
Kolmogorov–Smirnov	D	—	$Pr > D$	—
Cramer–von Mises	W-Sq	0.50616684	$Pr > W - Sq$	<0.010
Anderson–Darling	A-Sq	2.97049862	$Pr > A - Sq$	<0.010
	External Data			
Test	Statistic		p Value	
Kolmogorov–Smirnov	D	—	$Pr > D$	—
Cramer–von Mises	W-Sq	13.2377542	$Pr > W - Sq$	<0.010
Anderson–Darling	A-Sq	72.2392645	$Pr > A - Sq$	<0.010

Note: Kolmogorov–Smirnov not available for this model.

Table 7.15: *Quantiles for Weibull distribution in the guided example data*

Percent	Internal data		External data	
	Observed	Estimated	Observed	Estimated
1.0	0.003	0.00087	0.003	0.000003
5.0	0.021	0.00556	0.007	0.000127
10.0	0.030	0.01258	0.010	0.000648
25.0	0.049	0.03936	0.016	0.006300
50.0	0.081	0.10683	0.033	0.046159
75.0	0.201	0.23470	0.063	0.221823
90.0	0.390	0.41757	0.249	0.699945
95.0	0.780	0.56299	1.672	1.270266
99.0	1.773	0.91738	19.439	3.3633702

Table 7.16: *Fitted gamma distribution for the guided example data*

	Internal data	
Parameter	Symbol	Estimate
Threshold	Theta	0
Scale	Alpha	0.18815
Shape	Sigma	0.933084
Mean		0.17556
Std Dev		0.181746
	External data	
Parameter	Symbol	Estimate
Threshold	Theta	0
Scale	Alpha	2.670307
Shape	Sigma	0.25422
Mean		0.678844
Std Dev		1.346374

Note: SAS parametrization.

Table 7.17: *Quantiles for the gamma distribution in the guided example data*

Percent	Internal data		External data	
	Observed	Estimated	Observed	Estimated
1.0	0.003	0.00132	0.003	0.00000002
5.0	0.021	0.00753	0.007	0.00001378
10.0	0.030	0.01620	0.010	0.00021061
25.0	0.049	0.04695	0.016	0.00775882
50.0	0.081	0.11826	0.033	0.12265946
75.0	0.201	0.24338	0.063	0.71450826
90.0	0.390	0.41119	0.249	2.03510180
95.0	0.780	0.53901	1.672	3.26951301
99.0	1.773	0.83732	19.439	6.54785926

Table 7.18: *Fitted logistic distribution for the guided example data*

Internal data		
Parameter	Symbol	Estimate
Location	Mu	0.17556
Scale	Sigma	0.11521
External data		
Parameter	Symbol	Estimate
Location	Mu	0.67884
Scale	Sigma	2.25537

Table 7.19: *Fitted generalized Champernowne distribution*

Internal data		
Parameter	Symbol	Estimate
Shape	a	1.749957
Location	M	0.079500
Scale/threshold	c	0.000000
External data		
Parameter	Symbol	Estimate
Shape	a	1.270400
Location	M	0.033000
Scale/threshold	c	0.000000

7.2.1 *Fitting the Generalized Champernowne Distribution in SAS©*

We provide an SAS© program that can be used to fit the generalized Champernowne distribution both to the *internal data set* and the *external data set*. The parameter estimates for the generalized Champernowne distribution obtained in the *internal data set* and in the *external data set*, respectively, are presented in Table 7.19.

```
/*PARAMETRIC ESTIMATION Champernowne*/
proc iml;
 /*Read Internal and External data*/
  use oper.int_sim_table2_1 var {y};
  read all into x_int;
  use oper.ext_sim_table2_1 var {y};
  read all into x_ext;
 /*Modify the following line to select the
   Internal or the External data set*/
  x=x_int;
 /*nr is the number of observations in the sample*/
  nr=nrow(x);

 /*DEFINITION OF FUNCTIONS*/
 /*Median*/
    start median(v);
      n=nrow(v); r=rank(v); xsort=j(n,1,0);
      do i=1 to n;
        xsort[r[i]]=v[i];
      end;
      m=(xsort[n/2]+xsort[(n/2)+1])/2;
     return(m);
    finish;
 /*Champernowne CDF*/
    start Tcap(v,a,c,M);
      n=nrow(v);
      rr=1/2*(((v+j(n,1,c))##a-j(n,1,(M+c)**a))
         /((v+j(n,1,c))##a+j(n,1,(M+c)**a)
         -2*j(n,1,c**a))+1);
      return(rr);
    finish;

    *...(continues on the next page)...;
```

```
*...(continues from the previous page)...;
/*Likelihood*/
  start like(v,a,c,M),
    N=nrow(v);
    rr=N*log(a)+N*log((M+c)**a-c**a)
       +(a-1)*sum(log(v+j(n,1,c)))
       -2*sum(log((v+j(n,1,c))##a+j(n,1,(M+c)**a)
       -2*j(n,1,c**a)) );
    return(rr);
  finish;
/*DEFINITION OF VECTORS*/
par=j(3,1,0);
L=j(21,1,0);
aopt=j(1,1,0);
jmax=j(10,1,0);
cvalg=t(do(0,2,0.5));
ncvalg=nrow(cvalg);
avalg=j(ncvalg,1,0);
Lvalg=j(ncvalg,1,0);
/*FINDING THE OPTIMAL PARAMETERS FOR EVERY SAMPLE*/
M=median(x);
/*Finding the optimal a for every value of c*/
do k=1 to ncvalg;
  c=cvalg[k]*M;
  astart=0.1;
  aslut=21;
  do decimal=1 to 3;
    do j=1 to 21;
      a=astart+(j-1)*(aslut-astart)/20;
      L[j]=like(x,a,c,M);
end;
    jmax=loc(L=max(L));
    amax=astart+(jmax[1]-1)*(aslut-astart)/20;
    astart=max(0.0001,astart
          +(jmax[1]-2)*(aslut-astart)/20);
    aslut=astart+jmax[1]*(aslut-astart)/20;
  end;
      avalg[k]=amax;
      Lvalg[k]=L[jmax];
end;
/*Finding the optimal a and c for each sample*/
optk=loc(Lvalg=max(Lvalg));

*...(continues on the next page)...;
```

```
  *...(continues from the previous page)...;
  optk=optk[1];
  par[1]=avalg[optk];
  par[2]=cvalg[optk]*M;
  par[3]=M;
  Label={'a', 'c', 'M'};
  print Label par;
quit;
```

Next we provide an SAS© program that can be used to plot the probability distribution function (pdf) of the Champernowne distribution.

```
/*PLOTING PDF OF GENERALIZED CHAMPERNOWNE DISTRIBUTION*/
proc iml;
 /*Write the value of parameters for internal data*/
  a_I=1.749957;
  c_I=0;
  M_I=0.0795;
/*Write the value of parameters for external data*/
  a_E=1.2704;
  c_E=0;
  M_E=0.033;

 /*Define a grid*/
  x=t(do(0.3,8,0.001));
  nx=nrow(x);
 /*PDF of Champernowne distribution*/
  start pdf_ch(v,a,c,M);
   n=nrow(v);
   rr=a#(v+j(n,1,c))##(a-1)#((M+c)##a-c##a)/
      (((v+j(n,1,c))##a+(M+c)##a-2#c##a)##2);
   return(rr);
  finish;
  f_E=pdf_ch(x,a_E,c_E,M_E);
  f_I=pdf_ch(x,a_I,c_I,M_I);
  res=x||f_I||f_E;
/*Save density estimation for plot*/
 create oper.f_Champ from res [colname={'x' 'f_E' 'f_I'}];
  append from res;
 close oper.f_Champ;
quit;
 *...(continues on the next page)...;
```

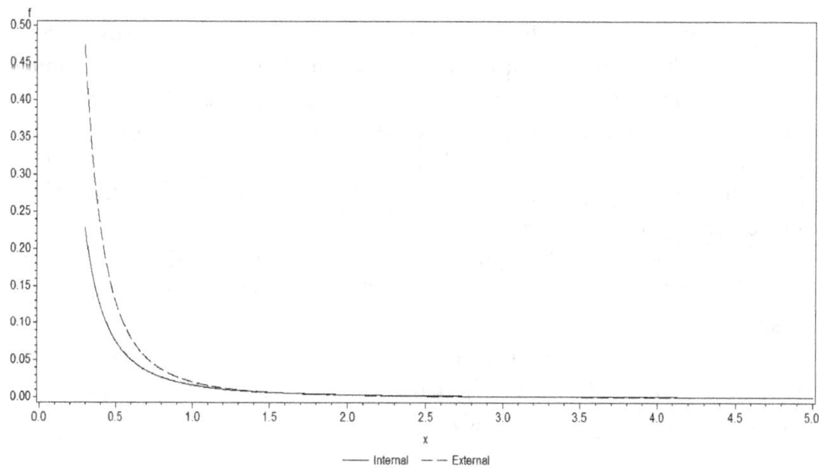

Figure 7.1: *Champernowne pdf for extreme internal and external data.*

```
 *...(continues from the previous page)...;
symbol1 interpol=join line=1 color=black;
symbol2 interpol=join line=2 color=black;
axis1 label=("x" justify=center) order=(0.5 to 5 by 0.5);
axis2 label=("f" justify=center) order=(0 to 0.15 by 0.01);
legend1 label=none value=("Internal" "External");
proc gplot data=oper.f_Champ;
  plot f_I*x=1 f_E*x=2 /overlay haxis=axis1 vaxis=axis2
                       legend=legend1;
run;
quit;
```

In Table 7.19 we can see that the shape parameter a for the *external data* is smaller than the corresponding shape parameter for the *internal data*. This means that *external data* distribution is heavier tailed than the one for the *internal data*. In Figure 7.1 we plot the estimated Champernowne pdf for both, the *internal* and the *external data* for values of losses above $VaR_{0.95}$. We observe that the tail for the *external data* is heavier.

7.2.2 *Quantile Estimation for Basic Parametric Distributions*

If a classical parametric distribution is fitted, then it is straightforward to obtain the estimated quantiles. Many functions have already been implemented in standard software programs, and they can be used right away. In the SAS© program, we present the implementation of quantile estimation for the logistic distribution, exponential distribution, gamma distribution, Weibull distribution, and lognormal distribution. We assume that the distribution parameters have previously been fitted.

We will show later how to implement the quantile estimation for the generalized Champernowne distribution. This will be presented in the next subsection.

The SAS© programs presented here allow us to obtain quantile estimation for a value of probability α (i.e., `alpha=0.995`) or for a grid (i.e., `alpha=t(do(0.95,0.999,0.001))`).

```
/*QUANTILE ESTIMATION WITH CLASSICAL
DISTRIBUTIONS WITH INTERNAL DATA*/
proc iml;
 /*Insert the probability level*/
  *alpha=0.995;

 /*Define a grid of probabilities*/
  alpha=t(do(0.95,0.999,0.001));
  na=nrow(alpha);

  /*MATRIX OF RESULTS*/
   VaRg=j(na,6,0);
   do i=1 to na;
    VaRg[i,1]=alpha[i];

  /*Logistic distribution*/
  /*Write the value of parameters*/
     mu=0.17556;
     sigma=0.1520532;
     VaR=quantile('LOGISTIC',alpha[i],mu,sigma);
     Label='LOGISTIC';
     if na=1 then
     print Label alpha VaR;
     VaRg[i,2]=VaR;

   *...(continues on the next page)...;
```

```
*...(continues from the previous page)...;

 /*Exponential distribution*/
 /*Write the value of parameters*/
  lambda=0.17556;
  VaR=quantile('EXPONENTIAL',alpha[i],lambda);
  Label='EXPONENTIAL';
  if na=1 then
  print Label alpha VaR;
  VaRg[i,3]=VaR;

 /*Gamma distribution*/
 /*Write the value of parameters*/
  a=0.933084;
  s=0.18815;
  VaR=quantile('GAMMA',alpha[i],a,s);
  Label='GAMMA';
  if na=1 then
  print Label alpha VaR;
  VaRg[i,4]=VaR;

 /*Weibull distribution*/
 /*Write the value of parameters*/
  a=0.88066;
  b=0.161968;
  VaR=quantile('WEIBULL',alpha[i],a,b);
  Label='WEIBULL';
  if na=1 then
  print Label alpha VaR;
  VaRg[i,5]=VaR;

 /*Lognormal distribution*/
 /*Write the value of parameters*/
  mu=-2.36353;
  sigma=1.06294;
  VaR=quantile('LOGNORMAL',alpha[i],mu,sigma);
  Label='LOGNORMAL';
  if na=1 then
  print Label alpha VaR;
  VaRg[i,6]=VaR;
end;

*...(continues on the next page)...;
```

```
  *...(continues from the previous page)...;

  /*Save quantile estimation for plot*/
  /*'alpha' is the variable grid values and 'q1' is
    logistic quantile, 'q2' is exponential, 'q3' gamma,
    'q4' Weibull and 'q5' lognormal*/

  create oper.q_classical from VaRg [colname={'alpha' 'q1'
                                  'q2' 'q3' 'q4' 'q5'}];
    append from VaRg;
  close oper.q_classical;
 quit;

 symbol1 interpol=join value=plus color=black;
 symbol2 interpol=join value=star color=black;
 symbol3 interpol=join value=square color=black;
 symbol4 interpol=join value=dot color=black;
 symbol5 interpol=join value=circle color=black;

 axis1 label=("alpha" justify=center)
       order=(0.95 to 1 by 0.01);

 axis2 label=("Quantile" justify=center )
       order=(0 to 3 by 0.1);

 legend1 label=none
       value=("Logistic" "Exponential" "Gamma"
              "Weibull" "Lognormal");

 proc gplot data=oper.q_classical;
  plot q1*alpha=1 q2*alpha=2 q3*alpha=3 q4*alpha=4
       q5*alpha=5/overlay haxis=axis1 vaxis=axis2
                  legend=legend1;
 run;
 quit;
```

The results obtained for the *internal data set* with the 99.5% quantile are presented in Table 7.20. In Figure 7.2 we plot the estimated quantile for a grid of probabilities α using the five classical distribution in Table 7.20. The curve can easily be obtained with standard plotting procedures.

Table 7.20 *Value-at-Risk (VaR) with tolerance level 99.5% for five distributions using the internal data set*

Distribution	Value-at-Risk 99.5%
Logistic	0.9804239
Exponential	0.9301726
Gamma	0.9662076
Weibull	1.0757134
Lognormal	1.4541506

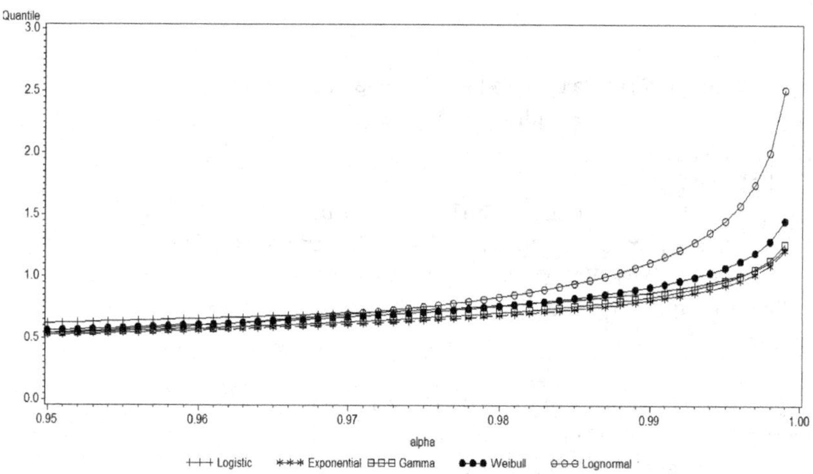

Figure 7.2 *Estimated quantile (VaR) with classical distributions for the internal data set.*

7.2.3 *Quantile Estimation for the Generalized Champernowne Distribution*

An SAS© program is given in order to show how the inverse of the distribution function of the generalized Champernowne distribution can be specified. Then the quantile can be calculated immediately.

```
/*QUANTILE ESTIMATION WITH CHAMPERNOWNE
  DISTRIBUTION FOR INTERNAL DATA*/
proc iml;
 /*Insert the probability*/
  *alpha=0.995;
 /*Define a grid of probabilities*/
  alpha=t(do(0.95,0.999,0.001));
  na=nrow(alpha);
  print na;
 /*MATRIX OF RESULTS*/
  VaRg=j(na,2,0);
  do i=1 to na;
   VaRg[i,1]=alpha[i];
   /*Champernowne distribution*/
   /*Write the value of parameters*/
    a=1.2704;
    c=0;
    M=0.033;
    VaRg[i,2]=((alpha[i]*(M+c)**a-(2*alpha[i]-1)*c**a)/
              (1-alpha[i]))**(1/a)-c;
  end;
  Label='CHAMPERNOWNE';
  if na=1 then print Label alpha VaR;
  create oper.q_champ from VaRg [colname={'alpha' 'q'}];
   append from VaRg;
  close oper.q_champ;
quit;

symbol6 interpol=join line=1 color=black;
axis1 label=("alpha" justify=center)
      order=(0.95 to 1 by 0.01);
axis2 label=("Quantile" justify=center )
      order=(0 to 8 by 0.5);
proc gplot data=oper.q_champ;
  plot q*alpha=6/haxis=axis1 vaxis=axis2;
run;
quit;
```

Table 7.21 *Value-at-Risk (VaR) with tolerance level 99.5% for the generalized Champernowne distribution using the internal data set*

Distribution	Value-at-Risk 99.5%
Champernowne	1.6369286

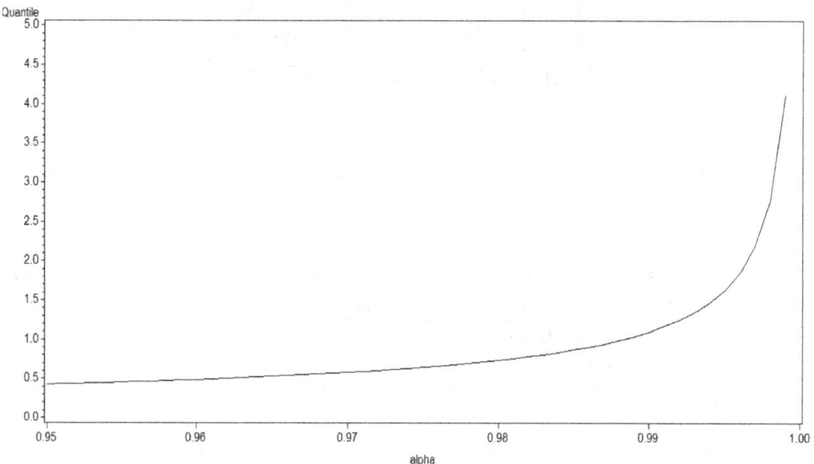

Figure 7.3 *Estimated quantile (VaR) with the Champernowne distribution for the internal data set.*

The results for the quantile at the 99.5% probability of a generalized Champernowne distribution for the *internal data set* are shown in Table 7.21. Later, in Figure 7.3, we plot the estimated quantile with the Champernowne distribution for the *internal data set*.

The corresponding 99.5% quantile results for the *external data set* are presented in Table 7.22, where all fitted classical distributions are presented, and in Table 7.23 we present the 99.5% estimated quantile with the Champernowne distribution for the *external data set*. In Figure 7.4 we plot the estimated quantiles for classical distributions, and in Figure 7.5 we plot the estimated quantile with the Champernowne distribution for the *external data set*.

Table 7.22: *Value-at-Risk (VaR) with tolerance level 99.5% using the external data set*

Distribution	Value-at-Risk 99.5%
Logistic	12.617179
Exponential	3.596731
Gamma	8.067583
Weibull	4.620828
Lognormal	2.759581

Table 7.23 *Value-at-Risk (VaR) with tolerance level 99.5% for the generalized Champernowne distribution using the external data set*

Distribution	Value-at-Risk 99.5%
Champernowne	2.128455

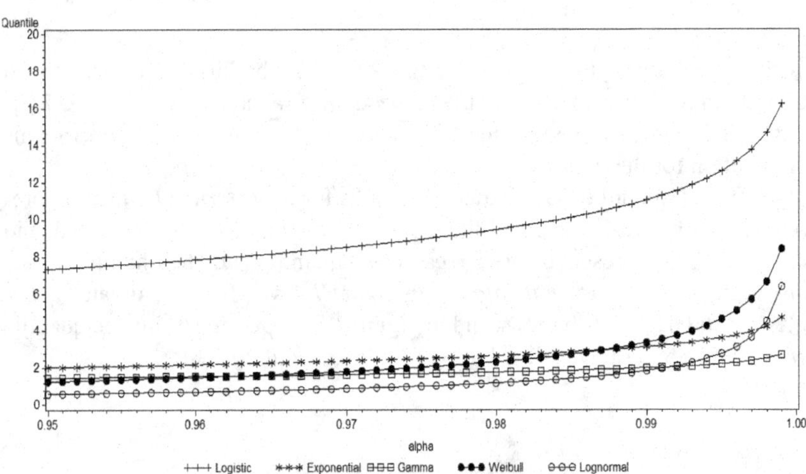

Figure 7.4 *Estimated quantile (VaR) with classical distributions for the external data set.*

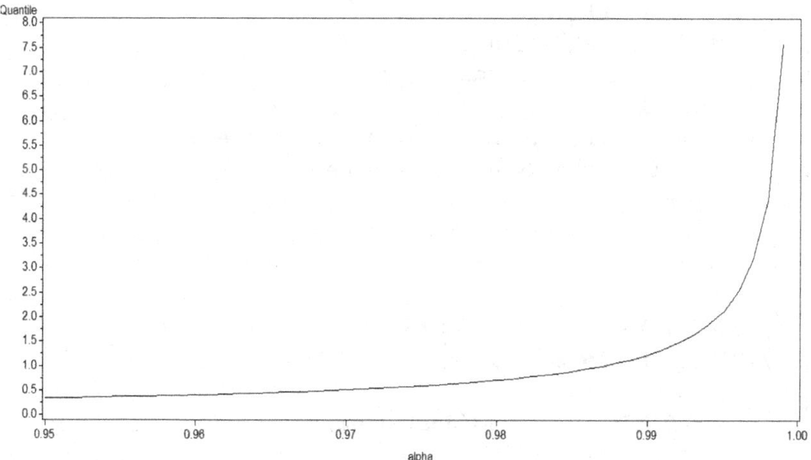

Figure 7.5 *Estimated quantile (VaR) with the Champernowne distribution for the external data set.*

7.3 Transformation Kernel Estimation

The program code implements the estimation method described in Chapter 3. It begins with classical kernel estimation and then uses the Champernowne transformation to apply the kernel transformation density estimation. The Epanechnikov kernel function is used. The results are stored, and the estimated densities can be plotted later.

```
/*CLASSICAL KERNEL DENSITY ESTIMATION*/

proc iml;
  /*Read Internal and External data*/
    use oper.int_data_table2_1;
     read all into x_int;
    use oper.ext_data_table2_1;
     read all into x_ext;
  /*Modify the following line to select the
   Internal or the External data set*/
    x=x_int;
  /*nr is the number of observations in the sample*/
    nr=nrow(x);

  /*DEFINITION OF FUNCTIONS*/
  /*Epanechnikov kernel function*/
    start K(u);
      rr=0.75*(1-(u#u))#(abs(u)<=1);
      return (rr);
    finish;

  /*Classical kernel estimation*/
    start fhatx(x,v,nr,b);
      /*x is a scalar, v is a vector*/
      rr=sum(K((x*j(nr,1,1)-v)/b))/(nr*b);
      return(rr);
    finish;

  /*Calculate median for grid value*/
    start median(x);
      n=nrow(x); r=rank(x); xsort=j(n,1,0);

    *...(continues on the next page)...;
```

```
    *...(continues from the previous page)...;

      do i=1 to n;
        xsort[r[i]]=x[i];
      end;
      m=(xsort[n/2]+xsort[(n/2)+1])/2;
      return(m);
    finish;
    /*GRID POINTS*/
      g=t(do(-0.9995,0.9995,0.001));
      ngrid=nrow(g);

    /*DEFINITION OF VECTORS*/
      fhatval=j(ngrid,1,0);

    /*Champernowne distribution parameters*/
      sigma=sqrt((sum((x-(sum(x)/nr)#j(nr,1,1))##2))
                /nr);
      b=2.34*sigma*(nr**(-1/5));

    /*Variable shifted grid points*/
      M=median(x);
      gv=M*(j(ngrid,1,1)+g)#((j(ngrid,1,1)-g)##(-1));
    /*Values for the kernel density in the
      variable shifted grid points*/
      do j=1 to ngrid;
        fhatval[j]=fhatx(gv[j],x,nr,b);
      end;
      result=gv||fhatval;

    /*Save the density estimation for plot*/
    /*'x' is the variable grid values
       and 'fhat' is classical kernel density estimate
       for each grid value of x*/
    create oper.classical_kernel from
          result [colname={'x' 'fhat'}];
    append from result;
    close oper.classical_kernel;
quit;
```

Many of the steps in the program for the classical kernel estimation are also required for the transformation kernel density estimation. So the two program codes have many parts in common.

```
/*TRANSFORMED KERNEL DENSITY ESTIMATION*/

proc iml;
  /*Read Internal and External data*/
   use oper.int_data_table2_1;
    read all into x_int;
   use oper.ext_data_table2_1;
    read all into x_ext;
  /*Modify the following line to select the
   Internal or the External data set*/
  x=x_int;
  /*nr is the number of observations in the sample*/
  nr=nrow(x);

  /*DEFINITION OF FUNCTIONS*/
  /*Transformation*/
    start Tcap(v,a,c,M);
      n=nrow(v);
      rr=1/2*(((v+j(n,1,c))##a-j(n,1,(M+c)**a))
               /((v+j(n,1,c))##a+j(n,1,(M+c)**a)
               -2*j(n,1,c**a))+1);
      return(rr);
    finish;

  /*Epanechnikov kernel function*/
    start K(u);
      rr=0.75*(1-(u#u))#(abs(u)<=1);
      return (rr);
    finish;

  /*Integral of the Epanechnikov kernel */
    start kint(u);
      rr= 0.25*(3-u*u)*u;
      return (rr);
    finish;

  /*Boundary correction*/
    start grkorr(x,b);
      upperlim=min(1,(1-x)/b);
      lowerlim=max(-1,-x/b);
      rr=kint(upperlim)-kint(lowerlim);
      return (rr);
    finish;

  *...(continues on the next page)...;
```

```
*...(continues from the previous page)...;

/*Kernel density with boundary correction
  for the transformed data*/
  start fhatz(z,v,nr,b);
    /*z is a scalar, v is a vector*/
    rr=sum(K((z*j(nr,1,1)-v)/b))/(nr*b*grkorr(z,b));
    return(rr);
  finish;

/*Kernel density for the original data*/
   start fhatx(x,v,nr,b,a,c,M);
     Tcapx=Tcap(x,a,c,M);
     density=a*(x+c)**(a-1)*((M+c)**a-c**a)/
     (((x+c)**a+(M+c)**a-2*c**a)**2);
     rr=fhatz(Tcapx,v,nr,b)*density;
     return(rr);
   finish;

/*GRID POINTS*/
   g=t(do(-0.9995,0.9995,0.001));
   ngrid=nrow(g);

/*DEFINITION OF VECTORS*/
   fhatval_t=j(ngrid,1,0);
   fhatval=j(ngrid,1,0);

/*CALCULATE THE ERROR OF ESTIMATES
   COMPARED TO THE TRUE DISTRIBUTION */
/*Parameters of Champernowne distribution*/
   a=0.4801188;
   c=0;
   M=0.011562; *This is the median of the data;

 /*Transformed data*/
    z=Tcap(x,a,c,M);
    sigma=sqrt((sum((z-(sum(z)/nr)#j(nr,1,1))##2))
               /nr);
    b=2.34*sigma*(nr**(-1/5));

 /*Variable shifted grid points*/
    gv=M*(j(ngrid,1,1)+g)#((j(ngrid,1,1)-g)##(-1));

 *...(continues on the next page)...;
```

```
    *...(continues from the previous page)...;

      /*Transformed variable shifted grid points*/
      gz=Tcap(gv,a,c,M);

      /*Values for the classical kernel density
        estimation with boundary of the transformed
        variable in the transformed variable
        shifted grid points*/
      do j=1 to ngrid;
        fhatval_t[j]=fhatz(gz[j],z,nr,b);
      end;

      /*Values for the transformed kernel density
        estimation of the original variable
        in the variable shifted grid points*/
      do j=1 to ngrid;
        fhatval[j]=fhatx(gv[j],z,nr,b,a,c,M);
      end;
      result=gz||fhatval_t||gv||fhatval;

      /*Save the density estimation for plot*/
      /*'z' is the transformed variable grid
        values and 'fhatz' is classical kernel
        density estimate for the transformed
        variable*/
      /*'x' is the original variable grid values
        and 'fhatx' is the transformed kernel
        density estimation for the original variable*/
      create oper.transformed_kernel
             from result
             [colname={'z' 'fhatz' 'x' 'fhatx'}];
      append from result;
      close oper.transformed_kernel;
quit;
```

Plots of the estimated densities can easily be obtained with the standard plotting procedures.

Figure 7.6 shows the results of the classical kernel estimation for the *internal data set*, which can be obtained from the SAS© program. Figure 7.7 plots the transformation kernel density estimated for the same data set. Note that the scale is not the same in both figures, but it can easily be customized with the given code.

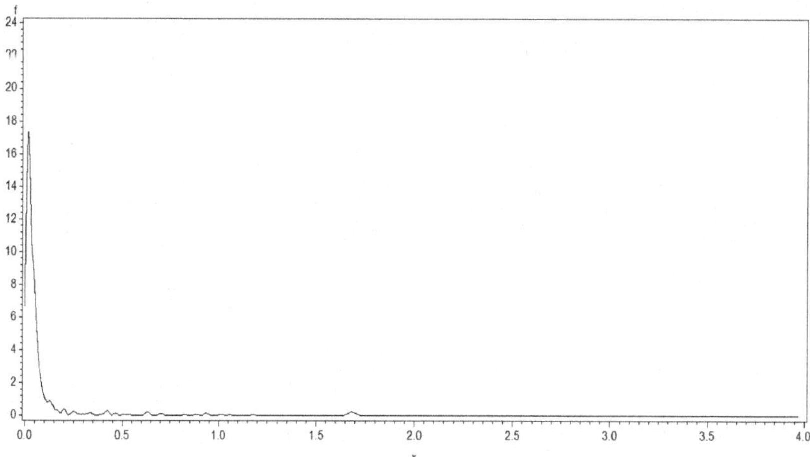

Figure 7.6: *Estimated classical kernel density for the internal data set.*

```
symbol1 interpol=join line=1;
axis1 label=("x" justify=center) order=(0 to 20 by 2);
axis2 label=("f" justify=center);
proc gplot data=oper.clasical_kernel;
  plot fhat*x=1/haxis=axis1 vaxis=axis2;
run;

symbol1 interpol=join line=1;
axis1 label=("x" justify=center) order=(0 to 5 by 0.5);
axis2 label=("f" justify=center) order=(0 to 1 by 0.1);
proc gplot data=oper.transformed_kernel;
  plot fhatx*x=1/haxis=axis1 vaxis=axis2;
run;
```

An interesting feature is that the transformation approach has smoothed down the bumps that were caused in the classical method by the fixed bandwidth in the classical kernel approach. We clearly do not recommend the classical kernel estimation procedure for heavy-tailed data because the classical method produces this kind of fluctuations in the density estimation around the extreme values of the tail.

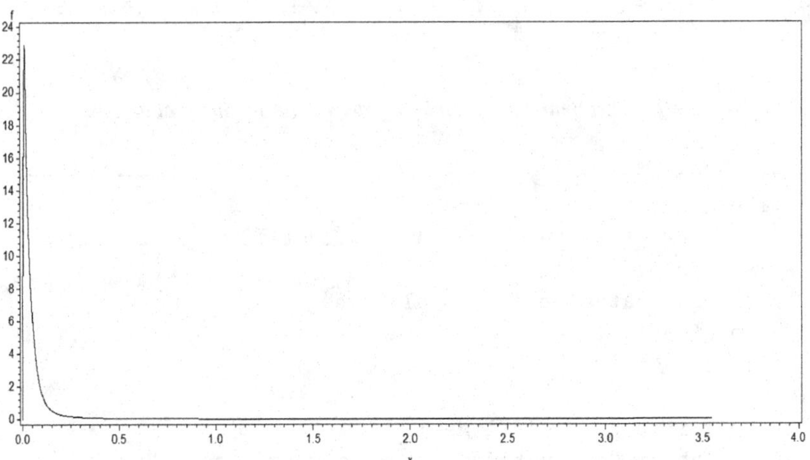

Figure 7.7: *Transformation kernel density estimate for the internal data set.*

7.4 Combining Internal and External Data

The programs presented in this section implement the methods developed in Chapter 4. We start with simulation of Poisson values with a given lambda parameter. This parameter is set to 30. To reproduce the exact same simulation, we mention that we have been working with SAS© version 9.2–Education Analytical Suite for operating system Windows 7. Also, in order to to initialize the simulation, we used a seed number equal to 100.

In the following program, for instance, if in one given run the Poisson generator has produced a value equal to 25, then we will assume that only 25 loss events occur. We fit a generalized Champernowne distribution again to our data set. Then we need to generate 25 random values from the fitted distribution.

```
/*SIMULATION OF YEARLY FREQUENCIES OF LOSSES*/
proc iml;
  R=10000;
  lambda=30;
  ran=uniform(j(R,1,100));
  lambda_r=quantile('POISSON',ran,lambda);
  create oper.lambda_r
    from lambda_r [colname={'l_r'}];
  append from lambda_r;
  close oper.lambda_r;
quit;
proc means data=oper.lambda_r;
 var l_r;
run;
proc univariate data=oper.lambda_r noprint;
 var l_r;
 histogram/midpoints = 11 12 13 14 15 16 17
                       18 19 20 21 22 23 24 25
                       26 27 28 29 30 31 32 33
                       34 35 36 37 38 39 40
                       41 42 43 44 45 46 47
                       48 49 50 51 52;
run;
```

In Table 7.24 we show the descriptive statistics obtained with the MEANS procedure, and in Figure 7.8 the histogram for the simulated yearly frequencies of losses is represented.

Table 7.24: *The Means procedure: output for our simulated data*

N	1000
Mean	30.0191
Std Deviation	5.5238
Minimum	11.0000
Maximum	52.0000

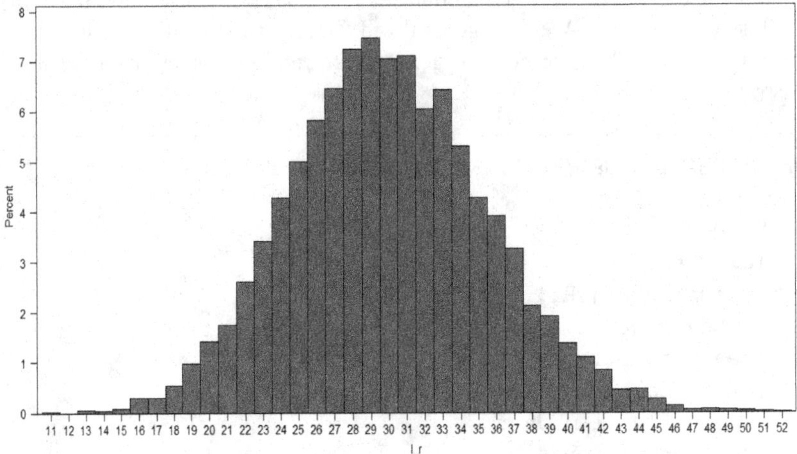

Figure 7.8: *Histogram for the simulated yearly frequencies of losses.*

Using the estimated parameter of the generalized Champernowne distribution with *internal* and *external data* that we show in Table 7.19, we simulated the five models proposed in Chapter 4. The generalized Champernowne transformation kernel estimation for internal data is used in the following program. This model was denoted by M1 in Chapter 4.

```
/*TRANSFORMED KERNEL ESTIMATION WITH INTERNAL DATA*/
/*Model M1 (only internal data)*/
proc iml;
 /*Read Internal data*/
   use oper.int_sim_table2_1 var {y};
   read all into x_int;

  *...(continues on the next page)...;
```

```
*...(continues from the previous page)...;

/*Read simulated frequencies */
  use oper.lambda_r;
  read all into lambda_r;
  x=x_int;

/*The number of simulated claims frequencies*/
  R=nrow(lambda_r);

/*nr is the number of observations in the sample*/
  nr=nrow(x);

/*DEFINITION OF FUNCTIONS*/
/*Transformation*/
  start Tcap(v,a,c,M);
    n=nrow(v);
    rr=1/2*(((v+j(n,1,c))##a-j(n,1,(M+c)**a))
          /((v+j(n,1,c))##a+j(n,1,(M+c)**a)
          -2*j(n,1,c**a))+1);
    return(rr);
  finish;

/*Epanechnikov Kernel function*/
  start K(u);
    rr=0.75*(1-(u#u))#(abs(u)<=1);
    return (rr);
  finish;

/*Integral of the Epanechnikov kernel*/
  start kint(u);
    rr= 0.25*(3-u*u)*u;
    return (rr);
  finish;

/*Boundary correction*/
   start grkorr(x,b);
     upperlim=min(1,(1-x)/b);
     lowerlim=max(-1,-x/b);
     rr=kint(upperlim)-kint(lowerlim);
     return (rr);
   finish;

*...(continues on the next page)...;
```

```
 *...(continues from the previous page)...;

/*Kernel density with boundary
  correction for the transformed data*/
  start fhatz(z,v,nr,b);
    /*z is a scalar, v is a vector*/
    rr=sum(K((z*j(nr,1,1)-v)/b))
       /(nr*b*grkorr(z,b));
    return(rr);
 finish;

/*Kernel density for the original data*/
 start fhatx(x,v,nr,b,a,c,M);
   Tcapx=Tcap(x,a,c,M);
   density=a*(x+c)**(a-1)*((M+c)**a-c**a)/
           (((x+c)**a+(M+c)**a-2*c**a)**2);
   rr=fhatz(Tcapx,v,nr,b)*density;
   return(rr);
 finish;

/*FIRST GRID POINTS*/
 Max_x=max(x);
 mx=int(Max_x)+10;
 print mx;

/*Define the precision*/
 precis=0.01;
 g=t(do(0,mx,precis));
 ngrid=nrow(g);

/*DEFINITION OF VECTOR*/
 sx=j(R,1,0);

/*Parameters of Champernowne distribution
 for internal data*/
   a=1.749957;
   c=0.1508618;
   M=0.0795; *This is de median
            of transformed data;
   do i=1 to R;
     u=uniform(j(lambda_r[i],1,100+i));
/*Transformed data*/
     z=Tcap(x,a,c,M);

 *...(continues on the next page)...;
```

```
      *...(continues from the previous page)...;

      sigma=sqrt((sum((z-(sum(z)/nr)#j(nr,1,1))##2))/nr);
      b=2.34*sigma*(nr**(-1/5));

     /*Calculate simulated internal losses*/
      gx=j(lambda_r[i],1,0);
      do j=1 to lambda_r[i];
        u2=fhatx(g[1],z,nr,b,a,c,M)*precis;
        do l=2 to ngrid until(u2>u[j]);
          u2=u2+fhatx(g[l],z,nr,b,a,c,M)*precis;
          gx[j]=g[l];
        end;
      end;
      sx[i]=sum(gx);
    end;

    create oper.SM1 from sx [colname={'s_r'}];
    append from sx;
    close oper.SM1;
quit;

axis1 order=(0 to 20 by 1);

proc univariate data=oper.SM1 noprint;
 var s_r;
 histogram/ vaxis=axis1 midpoints=0.50 1 1.50 2 2.50
                                  3.50 4 4.50 5 5.50
                                  6 6.50 7 7.50 8 8.50
                                  9 9.50 10 10.50 11
                                  11.50 12 12.5 13;
run;
quit;
```

The histogram for simulated operational loss with model M1 is shown in Figure 7.9.

Model M2, which uses the generalized Champernowne parameter estimates obtained with *external data*, is implemented in the SAS© program provided in this section.

Most parts of the program are exactly equal to the ones presented before. However, we reproduce it here again in order to show that only the fit, which is now obtained from the external data source, needs to be changed.

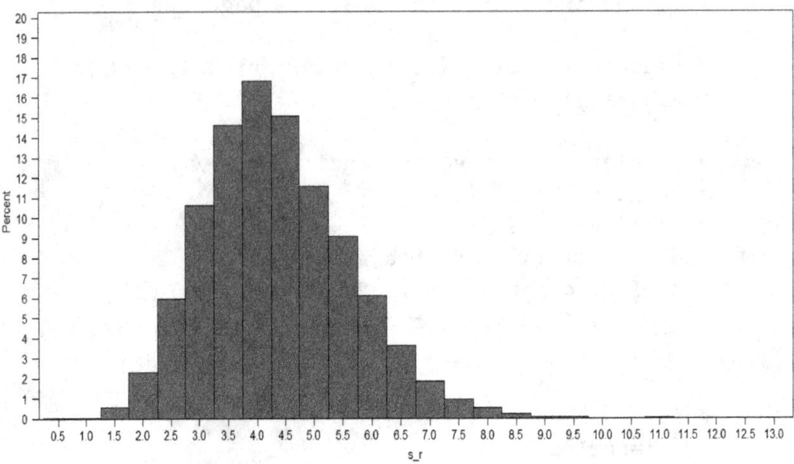

Figure 7.9: *Estimated total operational loss with model M1.*

```
/*Model M2 (all Champernowne parameters
  estimated with external data)*/
proc iml;
/*Read Internal data*/
  use oper.int_sim_table2_1 var {y};
  read all into x_int;
/*Read simulated frequencies */
  use oper.lambda_r;
  read all into lambda_r;
  x=x_int;
/*The number of simulated claims frequencies*/
  R=nrow(lambda_r);
/*nr is the number of observations in the sample*/
  nr=nrow(x);

/*DEFINITION OF FUNCTIONS*/
/*Transformation*/
   start Tcap(v,a,c,M);
     n=nrow(v);
     rr=1/2*(((v+j(n,1,c))##a-j(n,1,(M+c)**a))/
        ((v+j(n,1,c))##a+j(n,1,(M+c)**a)
        -2*j(n,1,c**a))+1);
     return(rr);
   finish;

 *...(continues on the next page)...;
```

```
*...(continues from the previous page)...;

/*Epanechnikov Kernel function*/
  start K(u);
    rr=0.75*(1-(u#u))#(abs(u)<=1);
    return (rr);
  finish;

/*Integral of the Epanechnikov kernel */
  start kint(u);
    rr= 0.25*(3-u*u)*u;
    return(rr);
  finish;

/*Boundary correction*/
  start grkorr(x,b);
    upperlim=min(1,(1-x)/b);
    lowerlim=max(-1,-x/b);
    rr=kint(upperlim)-kint(lowerlim);
    return(rr);
  finish;

/*Kernel density with boundary correction
  for the transformed data*/
  start fhatz(z,v,nr,b);
    /*z is a scalar, v is a vector*/
    rr=sum(K((z*j(nr,1,1)-v)/b))/(nr*b*grkorr(z,b));
    return(rr);
  finish;

/*Kernel density for the original data*/
  start fhatx(x,v,nr,b,a,c,M);
    Tcapx=Tcap(x,a,c,M);
    density=a*(x+c)**(a-1)*((M+c)**a-c**a)/
       (((x+c)**a+(M+c)**a-2*c**a)**2);
    rr=fhatz(Tcapx,v,nr,b)*density;
    return(rr);
  finish;

 *...(continues on the next page)...;
```

```
  *...(continues from the previous page)...;

/*FIRST GRID POINTS*/
  Max_x=max(x);
  mx=int(Max_x)+10;
  print mx;

/*Define the precision*/
  precis=0.01;
  g=t(do(0,mx,precis));
   ngrid=nrow(g);

/*DEFINITION OF VECTORS*/
  sx=j(R,1,0);

/*Parameters of Champernowne distribution
  for external data*/
  a=1.2704;
  c=0.74858;
  M=0.033; *This is the median
            of transformed data;

  do i=1 to R;
    u=uniform(j(lambda_r[i],1,100+i));
    /*Transformed data*/
     z=Tcap(x,a,c,M);
     sigma=sqrt((sum((z-(sum(z)/nr)#j(nr,1,1))##2))
              /nr);
     b=2.34*sigma*(nr**(-1/5));
    /*Calculate simulated internal losses*/
     gx=j(lambda_r[i],1,0);
     do j=1 to lambda_r[i];
       u2=fhatx(g[1],z,nr,b,a,c,M)*precis;
       do l=2 to ngrid until(u2>u[j]);
        u2=u2+fhatx(g[l],z,nr,b,a,c,M)*precis;
        gx[j]=g[l];
      end;
    end;
    sx[i]=sum(gx);
  end;

 *...(continues on the next page)...;
```

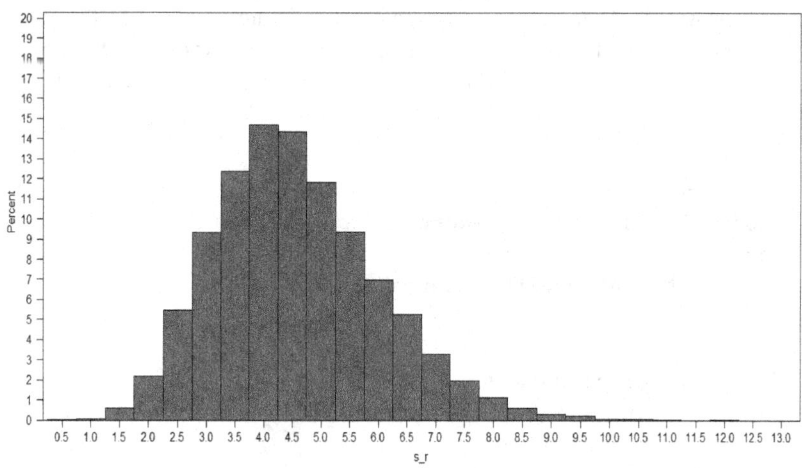

Figure 7.10: *Estimated total operational loss with model M2.*

```
   *...(continues from the previous page)...;

    create oper.SM2 from sx [colname={'s_r'}];
    append from sx;
    close oper.SM2;
quit;

axis1 order=(0 to 20 by 1);

proc univariate data=oper.SM2 noprint;
 var s_r;
 histogram/ vaxis=axis1 midpoints=0.50 1 1.50 2 2.50
                                  3.50 4 4.50 5 5.50
                                  6 6.50 7 7.50 8 8.50
                                  9 9.50 10 10.50 11
                                  11.50 12 12.5 13;
run;
quit;
```

The histogram for simulated operational loss with model M2 is shown in Figure 7.10. Note that the dispersion of the histogram is larger than the one observed in the previous section for model M1.

The program corresponding to model M3 in Chapter 4 is also given here. In this program, a combination of parameter estimates is used. In this case, the location parameter M is obtained from the *internal data set*, but the other parameters of the generalized Champernowne distribution are given by the *external data set*.

```
/*Model M3 (a and c Champernowne parameters
  estimated with external data
  and M estimated with internal data)*/

proc iml;
 /*Read Internal data*/
  use oper.int_sim_table2_1 var {y};
  read all into x_int;
 /* Read simulated frequencies */
  use oper.lambda_r;
  read all into lambda_r;
  x=x_int;

 /*The number of simulated claims frequencies*/
  R=nrow(lambda_r);

 /* nr is the number of observations in the sample*/
  nr=nrow(x);

 /*DEFINITION OF FUNCTIONS*/
 /*Transformation*/
    start Tcap(v,a,c,M);
      n=nrow(v);
      rr=1/2*(((v+j(n,1,c))##a-j(n,1,(M+c)**a))/
         ((v+j(n,1,c))##a+j(n,1,(M+c)**a)
         -2*j(n,1,c**a))+1);
      return(rr);
    finish;

 /*Epanechnikov Kernel function*/
    start K(u);
      rr=0.75*(1-(u#u))#(abs(u)<=1);
      return (rr);
    finish;

  *...(continues on the next page)...;
```

```
 *...(continues from the previous page)...;

/*Integral of the Epanechnikov kernel */
  start kint(u);
    rr= 0.25*(3-u*u)*u;
    return(rr);
  finish;

/*Boundary correction*/
  start grkorr(x,b);
    upperlim=min(1,(1-x)/b);
    lowerlim=max(-1,-x/b);
    rr=kint(upperlim)-kint(lowerlim);
    return(rr);
  finish;

/*Kernel density with boundary correction
  for the transformed data*/
  start fhatz(z,v,nr,b);
    /*z is a scalar, v is a vector*/
    rr=sum(K((z*j(nr,1,1)-v)/b))/(nr*b*grkorr(z,b));
    return(rr);
  finish;

/*Kernel density for the original data*/
  start fhatx(x,v,nr,b,a,c,M);
    Tcapx=Tcap(x,a,c,M);
    density=a*(x+c)**(a-1)*((M+c)**a-c**a)/
       (((x+c)**a+(M+c)**a-2*c**a)**2);
    rr=fhatz(Tcapx,v,nr,b)*density;
    return(rr);
  finish;

/*FIRST GRID POINTS*/
  Max_x=max(x);
  mx=int(Max_x)+10;
  print mx;
/*Define the precision*/
  precis=0.01;
  g=t(do(0,mx,precis));
  ngrid=nrow(g);

 *...(continues on the next page)...;
```

```
 *...(continues from the previous page)...;

/*DEFINITION OF VECTOR*/
  sx=j(R,1,0);

/*Parameters of Champernowne distribution
  for external and internal data*/
  a=1.2704;
  c=0.74858;
  M=0.0795; *This is the median
             of transformed internal data;

  do i=1 to R;
    u=uniform(j(lambda_r[i],1,100+i));

    /*Transformed data*/
    z=Tcap(x,a,c,M);
    sigma=sqrt((sum((z-(sum(z)/nr)#j(nr,1,1))##2))
               /nr);
    b=2.34*sigma*(nr**(-1/5));

    /*Calculate simulated internal losses*/
    gx=j(lambda_r[i],1,0);
    do j=1 to lambda_r[i];
      u2=fhatx(g[1],z,nr,b,a,c,M)*precis;
      do l=2 to ngrid until(u2>u[j]);
        u2=u2+fhatx(g[l],z,nr,b,a,c,M)*precis;
        gx[j]=g[l];
      end;
    end;
    sx[i]=sum(gx);
  end;

  create oper.SM3 from sx [colname={'s_r'}];
   append from sx;
  close oper.SM3;
quit;

 *...(continues on the next page)...;
```

```
   *...(continues from the previous page)...;

axis1 order=(0 to 20 by 1);

proc univariate data=oper.SM3 noprint;
 var s_r;
 histogram/ vaxis=axis1 midpoints=0.50 1 1.50 2 2.50
                                  3.50 4 4.50 5 5.50
                                  6 6.50 7 7.50 8 8.50
                                  9 9.50 10 10.50 11
                                  11.50 12 12.5 13;
run;
quit;
```

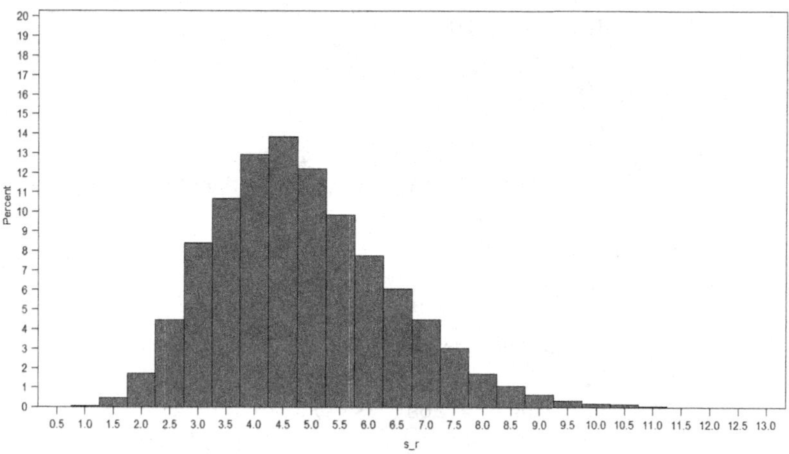

Figure 7.11: *Estimated total operational loss with model M3.*

The histogram for simulated operational loss with model M3 is show in Figure 7.11.

Model M4 has been developed in Chapter 4 and is also implemented in the SAS© code presented in this section. For this model only, the *external data set* has been used in order to produce the parameter estimates of the generalized Champernowne distribution that are used in the random variable generator.

Again, the core parts of the program are exactly the same as before, and only the distribution fit changes.

```
/*Model M4 (only external data)*/

proc iml;
  /*Read External data*/
  use oper.ext_sim_table2_1 var {y};
  read all into x_int;
  /* Read simulated frequencies */
  use oper.lambda_r;
  read all into lambda_r;
  x=x_int;

  /*The number of simulated claims frequencies*/
  R=nrow(lambda_r);

  /* nr is the number of observations in the sample*/
  nr=nrow(x);

  /*DEFINITION OF FUNCTIONS*/
  /*Transformation*/
    start Tcap(v,a,c,M);
      n=nrow(v);
      rr=1/2*(((v+j(n,1,c))##a-j(n,1,(M+c)**a))
            /((v+j(n,1,c))##a+j(n,1,(M+c)**a)
            -2*j(n,1,c**a))+1);
      return(rr);
    finish;

  /*Epanechnikov Kernel function*/
    start K(u);
      rr=0.75*(1-(u#u))#(abs(u)<=1);
      return (rr);
    finish;

  /*Integral of the Epanechnikov kernel*/
    start kint(u);
      rr= 0.25*(3-u*u)*u;
      return (rr);
    finish;

    *...(continues on the next page)...;
```

```
  *...(continues from the previous page)...;

/*Boundary correction*/
  start grkorr(x,b);
    upperlim=min(1,(1-x)/b);
    lowerlim=max(-1,-x/b);
    rr=kint(upperlim)-kint(lowerlim);
    return (rr);
  finish;

/*Kernel density with boundary
  correction for the transformed data*/
  start fhatz(z,v,nr,b);
    /*z is a scalar, v is a vector*/
    rr=sum(K((z*j(nr,1,1)-v)/b))
       /(nr*b*grkorr(z,b));
    return(rr);
  finish;

/*Kernel density for the original data*/
  start fhatx(x,v,nr,b,a,c,M);
    Tcapx=Tcap(x,a,c,M);
    density=a*(x+c)**(a-1)*((M+c)**a-c**a)/
            (((x+c)**a+(M+c)**a-2*c**a)**2);
    rr=fhatz(Tcapx,v,nr,b)*density;
    return(rr);
  finish;

/*FIRST GRID POINTS*/
  Max_x=max(x);
  mx=int(Max_x)+10;
  print mx;

/*Define the precision*/
  precis=0.01;
  g=t(do(0,mx,precis));
  ngrid=nrow(g);

/*DEFINITION OF VECTOR*/
  sx=j(R,1,0);

  *...(continues on the next page)...;
```

```
    *...(continues from the previous page)...;

  /*Parameters of Champernowne distribution
    for external data*/
    a=1.2704;
    c=0.74858;
    M=0.033; *This is the median
              of transformed data;
    do i=1 to R;
      u=uniform(j(lambda_r[i],1,100+i));
      /*Transformed data*/
      z=Tcap(x,a,c,M);
      sigma=sqrt((sum((z-(sum(z)/nr)#j(nr,1,1))##2))
                 /nr);
      b=2.34*sigma*(nr**(-1/5));
      /*Calculate simulated internal losses*/
      gx=j(lambda_r[i],1,0);
      do j=1 to lambda_r[i];
        u2=fhatx(g[1],z,nr,b,a,c,M)*precis;
        do l=2 to ngrid until(u2>u[j]);
          u2=u2+fhatx(g[l],z,nr,b,a,c,M)*precis;
          gx[j]=g[l];
        end;
      end;
      sx[i]=sum(gx);
    end;
    create oper.SM4 from sx [colname={'s_r'}];
    append from sx;
    close oper.SM4;
quit;

axis1 order=(0 to 20 by 1);
proc univariate data=oper.SM4 noprint;
 var s_r;
 histogram/ vaxis=axis1 midpoints=0.50 1 1.50 2 2.50
                                  3.50 4 4.50 5 5.50
                                  6 6.50 7 7.50 8 8.50
                                  9 9.50 10 10.50 11
                                  11.50 12 12.5 13;
run;
quit;
```

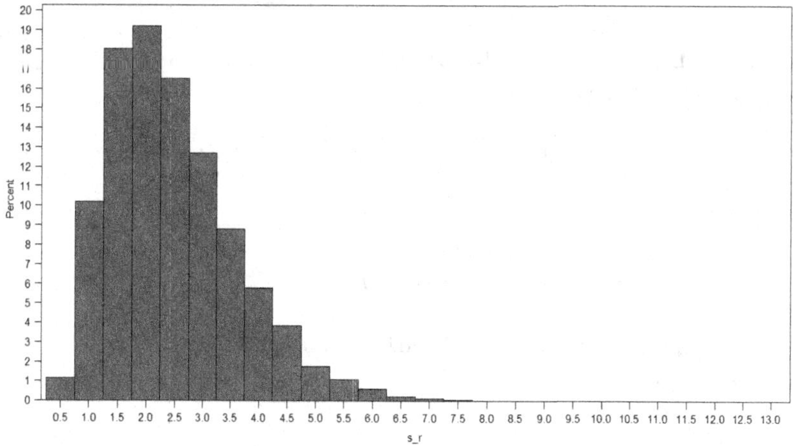

Figure 7.12: *Estimated total operational loss with model M4.*

The histogram for simulated operational loss with model M4 is shown in Figure 7.12.

In Chapter 4 we presented a case where we assumed a Weibull distribution for the *internal data set*. The implementation is shown in the SAS© program for this section.

It is straightforward to change model M5, which is the one here based on the Weibull distribution, by another model with another distribution.

```
/*Model M5 (Weibull assumption
  for internal data)*/

proc iml;
  /*Read Internal data*/
    use oper.int_sim_table2_1 var {y};
    read all into x_int;
  /*Read simulated frequencies */
  use oper.lambda_r;
  read all into lambda_r;
  x=x_int;

  *...(continues on the next page)...;
```

```
   *...(continues from the previous page)...;

  /*The number of simulated claims frequencies*/
    R=nrow(lambda_r);

  /* nr is the number of observations in the sample*/
  nr=nrow(x);

  /*Weibull distribution*/
  /*Write the value of parameters*/
    sx=j(R,1,0);
    /*Parameters of Weibull distribution*/
    a=0.88066;
    b=0.161968;
    do i=1 to R;
      u=uniform(j(lambda_r[i],1,100+i));
      /*Calculate simulated internal losses*/
      gx=j(lambda_r[i],1,0);
      do j=1 to lambda_r[i];
        gx[j]=quantile('WEIBULL',u[j],a,b);
      end;
      sx[i]=sum(gx);
    end;
    create oper.SM5 from sx [colname={'s_r'}];
    append from sx;
    close oper.SM5;
quit;
```

Finally, we obtain the histogram for the simulated operational loss with model M5, and the results are show in Figure 7.13.

The following program summarizes the results obtained in the previous codes when implementing the methods described in Chapter 4. It then produces an output that is similar to Table 4.2, but now we use the *internal data set* and the *external data set* that have been studied in the current guided example chapter and not the data that were discussed in Chapter 4.

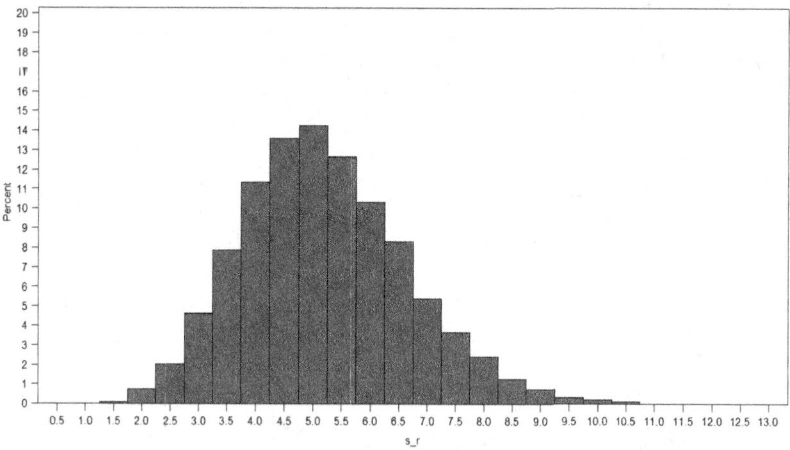

Figure 7.13: *Estimated total operational loss with model M5.*

```
proc iml;

  /*Read results on sum of claims*/
  use oper.SM1;
  read all into sm1;
  use oper.SM2;
  read all into sm2;
  use oper.SM3;
  read all into sm3;
  use oper.SM4;
  read all into sm4;
  use oper.SM5;
  read all into sm5;

  sm=sm1||sm2||sm3||sm4||sm5;

  /*nr is the number of simulations*/
  nr=nrow(xx);

  nc=ncol(sm);

  *...(continues on the next page)...;
```

```
*...(continues from the previous page)...;

/*STATISTICS*/
 /*Mean*/
  start mean(v);
    n=nrow(v);
    m=sum(v)/n;
    return(m);
  finish;

 /*Sd*/
  start sd(v);
    n=nrow(v);
    m=mean(v);
    s=sum((v-m)##2)/n;
    return(s);
  finish;

 /*Median*/
  start median(v);
    n=nrow(v);
    r=rank(v);
    xsort=j(n,1,0);
    do i=1 to n;
      xsort[r[i]]=v[i];
    end;
    m=(xsort[n/2]+xsort[(n/2)+1])/2;
    return(m);
  finish;

 /*VaR*/
  start Var(v,q);
    n=nrow(v);
    r=rank(v);
    xsort=j(n,1,0);
    do i=1 to n;
      xsort[r[i]]=v[i];
    end;
    m=(xsort[q*n]);
    return(m);
  finish;

*...(continues on the next page)...;
```

```
*...(continues from the previous page)...;

/*TVaR*/
start TVar(v,q);
  n=nrow(v);
  r=rank(v);
  xsort=j(n,1,0);
  do i=1 to n;
    xsort[r[i]]=v[i];
  end;
  m=Var(v,q);
  in=(xsort>=m);
  v=sum(in#xsort)/(n-q*n);
  return(v);
finish;

/*Results in Table 4.2*/
Table=j(5,9,0);

/*Mean*/
do i=1 to nc;
  s=sm[,i];
  Table[i,1]=mean(s);
end;

/*Median*/
do i=1 to nc;
  s=sm[,i];
  Table[i,2]=median(s);
end;

/*Sd*/
do i=1 to nc;
  s=sm[,i];
  Table[i,3]=sd(s);
end;

/*VaR95*/
do i=1 to nc;
  s=sm[,i];
  Table[i,4]=Var(s,0.95);
end;

*...(continues on the next page)...;
```

```
    *...(continues from the previous page)...;

    /*VaR99*/
    do i=1 to nc;
      s=sm[,i];
      Table[i,5]=Var(s,0.99);
    end;

    /*VaR999*/
    do i=1 to nc;
      s=sm[,i];
      Table[i,6]=Var(s,0.999);
    end;

    /*TVaR95*/
    do i=1 to nc;
      s=sm[,i];
      Table[i,7]=TVar(s,0.95);
    end;

    /*TVaR99*/
    do i=1 to nc;
      s=sm[,i];
      Table[i,8]=TVar(s,0.99);
    end;

    /*TVaR999*/
    do i=1 to nc;
      s=sm[,i];
      Table[i,9]=TVar(s,0.999);
    end;

    Mlabel={'M1','M2','M3','M4','M5'};
    Collabel={'Model' 'Mean' 'Median' 'Sd'
              'VaR95' 'VaR99' 'VaR999'
              'TVaR95' 'TVaR99' 'TVaR999'};
    print Collabel;
    print Mlabel Table;
quit;
```

In Table 7.25 we show the results for the *internal data set* when combined with the *external data set* of the guided example using the methods that have been presented in Chapter 4. Table 7.25 presents the mean, median, and stan-

Table 7.25 *General statistics for the five models from Chapter 4 with the guided example data sets*

Model	Mean	Median	Sd
M1	4.330906	4.22	1.5416004
M2	4.568492	4.42	2.0063268
M3	4.823747	4.65	2.3676564
M4	2.469918	2.28	1.2398481
M5	5.204205	5.08	2.0720341

Table 7.26 *Quantile statistics for the five models from Chapter 4 with the guided example data sets*

Model	VaR 95%	VaR 99%	VaR 99.9%
M1	6.53	7.65	8.84
M2	7.11	8.38	9.89
M3	7.60	8.93	10.68
M4	4.56	5.74	7.16
M5	7.76	8.99	10.37

dard deviation for the five models, Table 7.26 shows the VaR, and Table 7.27 shows the Tail-Value-at-Risk (TVaR) for these models.

As expected, model M4 provides lower risk measures, while models M3 and M5 have larger risk measures than the rest of models. It seems that the use of the Weibull distributional assumption leads to larger risk values than those obtained with other procedures.

7.5 Underreporting Implementation

The programs presented in this section show how to implement underreporting information, as described in Chapter 5. Some parts of the program have been used to implement the previous chapters, but they are repeated again so that the program content is complete.

Table 7.27 *Tail-Value-at-Risk for the five models from Chapter 4 with the guided example data sets*

Model	TVaR 95%	TVaR 99%	TVaR 99.9%
M1	7.24	8.27	10.23
M2	7.90	9.16	11.45
M3	8.50	10.06	12.39
M4	5.28	6.40	8.36
M5	8.52	9.72	11.89

In order to implement the methods shown in Chapter 5, we use two data sets that emulate external publicly available data. They are called c1_sim_table5_2 for *public data on risk no. 1*, and c2_sim_table5_2 for *public data on risk no. 2*.

```
/*SIMULATION OF ANNUAL FREQUENCIES
OF LOSSES FOR EACH RISK CATEGORY*/
proc iml;
  /*R is the number of repetitions in the
     simulation */
   R=10000;
  /*Frequency in each risk category (scenario data)*/
  lambda_i={10,20};

  /*Number of risk categories*/
   ncat=nrow(lambda_i);
   lambda_ri=j(R,ncat,0);
   do i=1 to ncat;
     ran=uniform(j(R,1,99+i));
     lambda_ri[,i]=quantile('POISSON',ran,lambda_i[i]);
   end;
   res=ran||lambda_ri;
 /*Data base with simulated number of claims*/
   create oper.lambda_ri
          from res [colname={'ran' 'l_r1' 'l_r2'}];
   append from res;
   close oper.lambda_ri;
quit;

axis1 order=(0 to 14 by 1);
proc univariate data=oper.lambda_ri noprint;
 var l_r1;
 histogram/vaxis=axis1 midpoints=1 2 3 4 5 6 7 8 9 10 11
                                 12 13 14 15 16 17
                                 18 19 20 21 22 23;
run;
proc univariate data=oper.lambda_ri noprint;
 var l_r2;
 histogram/vaxis=axis1 midpoints=5 6 7 8 9 10 11 12 13 14
                                 15 16 17 18 19 20 21 22 23
                                 24 25 26 27 28 29 30 31 32
                                 33 34 35 36 37 38 39;
run;
```

For simulated annual frequencies of losses for each risk category, we use random simulation seed numbers 100 and 101, respectively. In Figures 7.14 and 7.15 we show, respectively, the histograms of simulated annual frequencies losses for each category.

```
/*INTERPOLATION OF UNDERREPORTING FUNCTION*/
proc iml;
/*Report level*/
 val=0//0.001//0.010//0.100//0.500//1//10//50//100;
 mx=max(val);/*Define maximal for grid points*/
 ncat=2; /*Number of risk categories*/

/*Grid points*/
 precis=0.01;
 g=t(do(0,mx,precis));
 ngrid=nrow(g);

/*Underreporting information*/
 u1={0,0.01,0.1,0.42,0.55,0.6,0.85,0.93,0.99};
 u2={0,0.05,0.1,0.32,0.43,0.5,0.73,0.88,0.99};
 u=u1||u2;
 nint=nrow(u);
 ui=j(ngrid,ncat,0);
 do i=1 to ncat;
 /*Interpolated underreporting function in grid points*/
  nin0=1;
  nin=0;
  do k=2 to nint;
    in1=(g>val[k]);
in2=(g>val[k-1]);
in=sum(in2)-sum(in1);
    nin=nin0+in;
    do j=nin0+1 to nin+1;
        ui[j,i]=u[k,i]-((val[k]-g[j])*(u[k,i]-u[k-1,i]))
                (val[k]-val[k-1]);
    end;
    nin0=nin;
  end;
 end;
 res=g||ui;
 res=res[2:ngrid,];

  *...(continues on the next page)...;
```

```
    *...(continues from the previous page)...;

/*Data base with interpolated underreporting functions*/
create oper.underrep from res [colname={'x' 'u1' 'u2'}];
  append from res;
close oper.underrep;
quit;

symbol1 interpol=join line=1 color=black;
symbol2 interpol=join line=2 color=black;
axis1 label=("Report Level" justify=center);
axis2 label=("u");
proc gplot data=oper.underrep;
  plot u1*x=1 u2*x=2 /overlay haxis=axis1 vaxis=axis2;
run;
quit;
```

In Figure 7.16 we show the underreporting function for each risk category. We note that the function increases with the cost, that is, the lower is the loss the higher is the probability of underreporting.

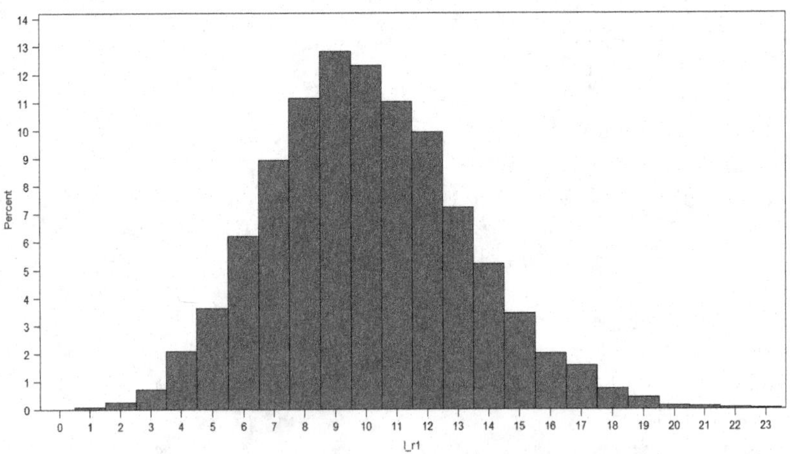

Figure 7.14: *Histogram for the simulated frequencies of losses for risk category 1.*

In Table 7.28 we show the parameter estimates of the Weibull distribution for risk categories 1 and 2. Once the Weibull model has been obtained, we follow the presentation made in Chapter 5. All models mentioned in that chapter, that is, WE, WE.UR, WE.KS, and WE.KS.UR, can be calculated with a single program, but here we have divided the program in parts so that the histograms of the results for each model can be shown immediately after they have been obtained.

```
/*WEIBULL PARAMETRIC ESTIMATION FOR EACH RISK CATEGORY*/
proc univariate data=oper.c1_sim_table5_2 noprint;
  var y;
  histogram / weibull;
run;
proc univariate data=oper.c2_sim_table5_2 noprint;
  var y;
  histogram / weibull;
  run;
```

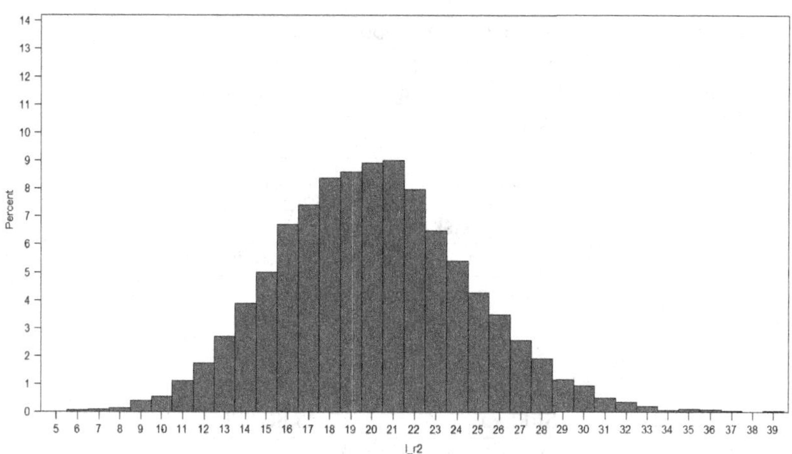

Figure 7.15: *Histogram for the simulated frequencies of losses for risk category 2.*

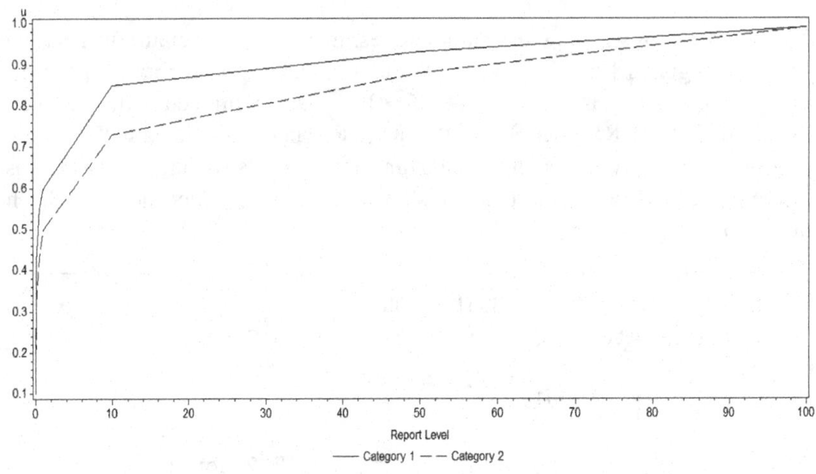

Figure 7.16: *Underreporting function for each risk category.*

Table 7.28: *Fitted Weibull distribution for our example of publicly available data*

Risk category 1		
Parameter	Symbol	Estimate
Threshold	Theta	0
Scale	Sigma	11.59642
Shape	C	0.51607
Mean		21.91461
Std Dev		46.90245
Risk category 2		
Parameter	Symbol	Estimate
Threshold	Theta	0
Scale	Sigma	10.48704
Shape	C	0.61704
Mean		15.22842
Std Dev		25.86213

Note: SAS parametrization.

```
/*RESULTS WITH WEIBULL*/

proc iml;

  /*Read data of risk category no 1.*/
  use oper.c1_sim_table5_2 var {y};
  read all into x_c1;

  /*Read data of risk category no 2.*/
  use oper.c2_sim_table5_2 var {y};
  read all into x_c2;

  /*Read simulated number of claims*/
  use oper.lambda_ri;
  read all into rl;
  ran=rl[,1];
  lambda_ri=rl[,2:3];

  /*Read Underreporting function*/
  use oper.underrep;
  read all into under;
  x=under[,1];
  und=under[,2:3];
  nx=nrow(x);
  precis=0.01;

  /*Number of claims in each category*/
  n1=nrow(x_c1);
  n2=nrow(x_c2);
  n=n1//n2;

  /*Concatenate cost of claims for each category*/
  xall=x_c1//x_c2;
  /*Number of risk categories*/
  ncat=2;

  *...(continues on the next page)...;
```

```
*...(continues from the previous page)...;

/*MODEL WE*/
/*Parameters of Weibull distribution*/

  Par_weib={0.516068 0.617035,
            11.59642 10.48704};
  print Par_weib;

/*The number of simulated claims*/
  R=nrow(lambda_ri);
  ngrid=nrow(x);
  sx=j(R,ncat,0);

  /*Total cost results without underreporting*/
  do k=1 to ncat;
    a=Par_weib[1,k];
    s=Par_weib[2,k];
    do i=1 to R;
      if lambda_ri[i,k]>0 then do;
        u=uniform(j(lambda_ri[i,k],1,100+i));
        /*Calculate simulated internal losses*/
        gx=j(lambda_ri[i,k],1,0);
        do j=1 to lambda_ri[i,k];
          gx[j]=quantile('WEIBULL',u[j],a,s);
        end;
        sx[i,k]=sum(gx);
      end;
    end;
  end;
  totsx=sx[,+];
  res=sx||totsx;
  /*Data base with total cost*/
  create oper.WE from res
               [colname={'s_1' 's_2' 'Total'}];
  append from res;
  close oper.WE;
*...(continues with the model WE.UR)...;
```

We obtain the histograms for simulated operational losses for each risk category and for the total operational losses. All histograms are shown in Figures 7.17, 7.18, and 7.19. We also present the histograms of the operational loss for each risk category and for the sum of the two, assuming that individual losses are generated from a Weibull distribution and without underreporting.

```
axis1 order=(0 to 44 by 2);

proc univariate data=oper.WE noprint;
 var s_1;
 histogram /vaxis=axis1
           midpoints=30 90 150 210 270 330 390 450
                     510 570 630 690 750 810 870
                     930 990 1050 1110 1170 1230
                     1290 1350 1410 1470 1530 1590
                     1650 1710 1770;
run;

axis1 order=(0 to 30 by 2);

proc univariate data=oper.WE noprint;
 var s_2;
 histogram/vaxis=axis1
          midpoints=20 60 100 140 180 220 260 300 340
                    380 420 460 500 540 580 620 660
                    700 740 780 820 860 900 940 980
                    1020 1060 1100 1140;
 run;

axis1 order=(0 to 34 by 2);

proc univariate data=oper.WE noprint;
 var Total;
 histogram/vaxis=axis1
               midpoints=90 180 270 360 450 540 630
                         720 810 900 990 1080 1170
                         1260 1350 1440 1530 1620 1710
                         1800 1890 1980 2070;
 run;
```

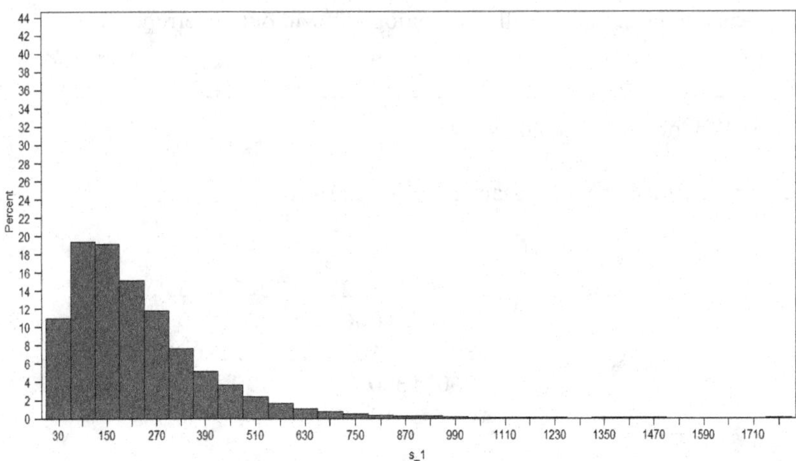

Figure 7.17 *Histogram of operational loss in our example of publicly available data for risk category 1: Weibull model without underreporting.*

```
*...(continues from the model WE)...;

/*MODEL WE.UR*/
/*Density of reported claims*/
f_R=j(nx,ncat,0);
do k=1 to ncat;
  a=Par_weib[1,k];
  s=Par_weib[2,k];
  f_R[,k]=pdf('WEIBULL',x,a,s);
end;

/*Density of occurred claims*/
f_O=j(nx,ncat,0);
precis=0.01;
do k=1 to ncat;
  /*Numerator*/
  num=f_R[,k]/und[,k];
  /*Denominator*/
  den=sum((f_R[,k]/und[,k])#precis);
  f_O[,k]=num/den;
end;

*...(continues on the next page)...;
```

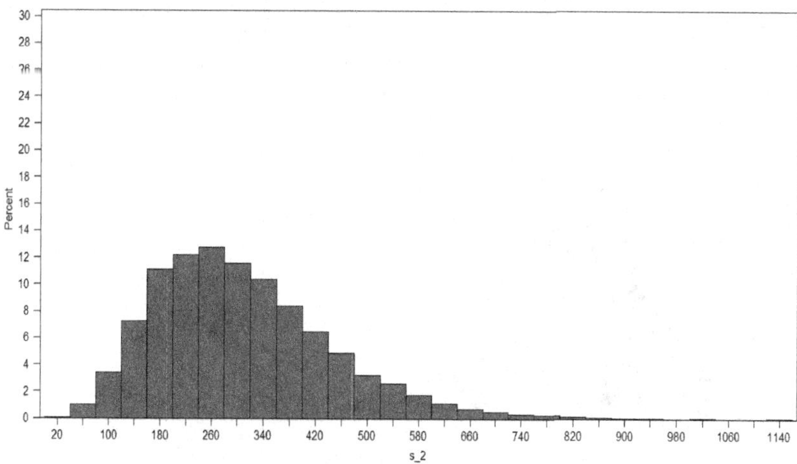

Figure 7.18 *Histogram of operational loss in our example of publicly available data for risk category 2: Weibull model without underreporting.*

```
*...(continues from the previous page)...;

/*Probability of reporting a claim*/
P_g=j(1,ncat,0);
do k=1 to ncat;
  P_g[k]=sum((f_O[,k]#und[,k])#precis);
end;
print P_g;

/*Number of occurred claims*/
lambda_i={10,20};
/*Frequency in each reported claims risk category*/
print lambda_i;
 lambda_O=lambda_i/P_g`;
/*Frequency in each occurred claims risk category*/
print lambda_O; /*Results Table 2 file We.UR*/
lambda_oi=j(R,ncat,0);

do k=1 to ncat;
  lambda_oi[,k]=quantile('POISSON',ran,lambda_o[k]);
end;

*...(continues on the next page)...;
```

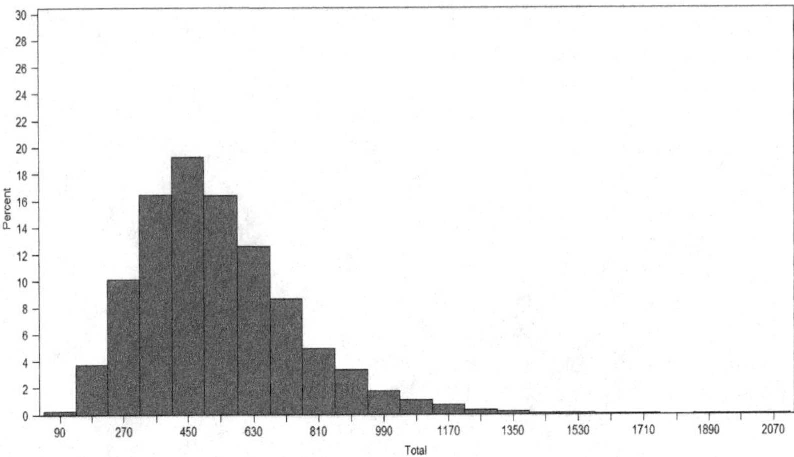

Figure 7.19 *Histogram of total operational loss in our example of publicly available data: Weibull model without underreporting.*

```
*...(continues from the previous page)...;

/*Total cost results with underreporting*/
sx=j(R,ncat,0);
do k=1 to ncat;
  do i=1 to R;
    if lambda_oi[i,k]>0 then do;
      u=uniform(j(lambda_oi[i,k],1,100+i));
      /*Calculate simulated internal losses*/
      gx=j(lambda_oi[i,k],1,0);
      do j=1 to lambda_oi[i,k];
        u2=f_O[1,k]*precis;
        do l=2 to nx until(u2>u[j]);
          u2=u2+f_O[l,k]*precis;
          gx[j]=x[l];
        end;
      end;
      sx[i,k]=sum(gx);
    end;
  end;
end;
totsx=sx[,+];
res=sx||totsx;

*...(continues on the next page)...;
```

```
*...(continues from the previous page)...;

/*Data set with total cost*/
create oper.WE_UR
      from res [colname={'s_1' 's_2' 'Total'}];
append from res;
close oper.WE_UR;

*...(continues with the model WE.KS)...;
```

In Figures 7.20, 7.21, and 7.22 we show histograms of operational loss for each risk category and for their sum, assuming that individual losses are generated for a Weibull distribution and with underreporting. When introducing underreporting, the frequency of small events increases substantially as a result of introducing the information that have originally not been reported.

```
axis1 order=(0 to 44 by 2);

proc univariate data=oper.WE_UR noprint;
 var s_1;
 histogram /vaxis=axis1
           midpoints=30 90 150 210 270 330 390 450
                     510 570 630 690 750 810 870
                     930 990 1050 1110 1170 1230
                     1290 1350 1410 1470 1530 1590
                     1650 1710 1770;
 run;

 axis1 order=(0 to 30 by 2);

 proc univariate data=oper.WE_UR noprint;
  var s_2;
  histogram/vaxis=axis1
           midpoints=20 60 100 140 180 220 260 300 340
                     380 420 460 500 540 580 620 660
                     700 740 780 820 860 900 940 980
                     1020 1060 1100 1140;
 run;

*...(continues on the next page)...;
```

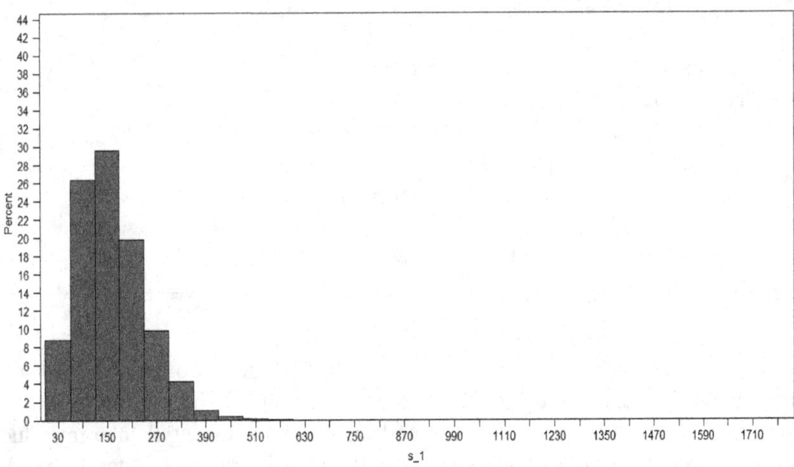

Figure 7.20 *Histogram of operational loss in our example of publicly available data for risk category 1: Weibull model with underreporting.*

```
*...(continues from the previous page)...;

axis1 order=(0 to 34 by 2);

proc univariate data=oper.WE_UR noprint;
 var Total;
 histogram/vaxis=axis1
           midpoints=90 180 270 360 450 540 630 720 810
                     900 990 1080 1170 1260 1350 1440
                     1530 1620 1710 1800 1890 1980 2070;
 run;
```

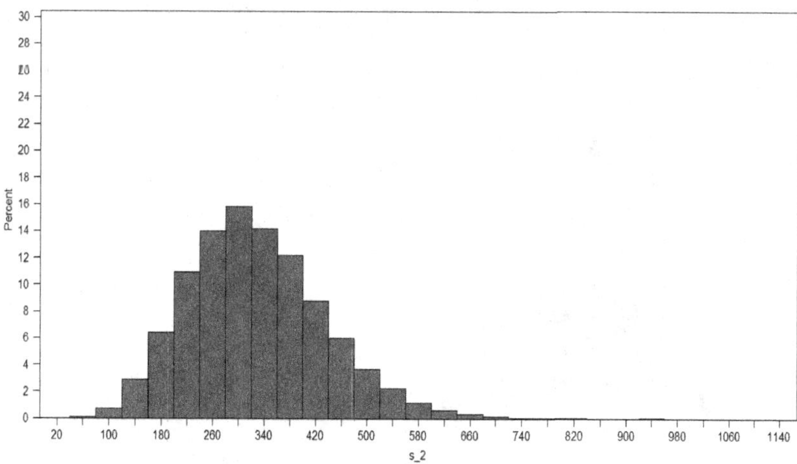

Figure 7.21 *Histogram of operational loss in our example of publicly available data for risk category 2: Weibull model with underreporting.*

```
*...(continues from the model WE.UR)...;

/*MODEL WE.KS*/

/*Weibull Transformation*/
  start Tcap(v,a,s);
    rr=cdf('WEIBULL',v,a,s);
    return(rr);
  finish;

/*Epanechnikov Kernel function*/
  start K(u);
    rr=0.75*(1-(u#u))#(abs(u)<=1);
    return (rr);
  finish;

/*Integral of the Epanechnikov kernel */
  start kint(u);
    rr= 0.25*(3-u*u)*u;
    return (rr);
  finish;

*...(continues on the next page)...;
```

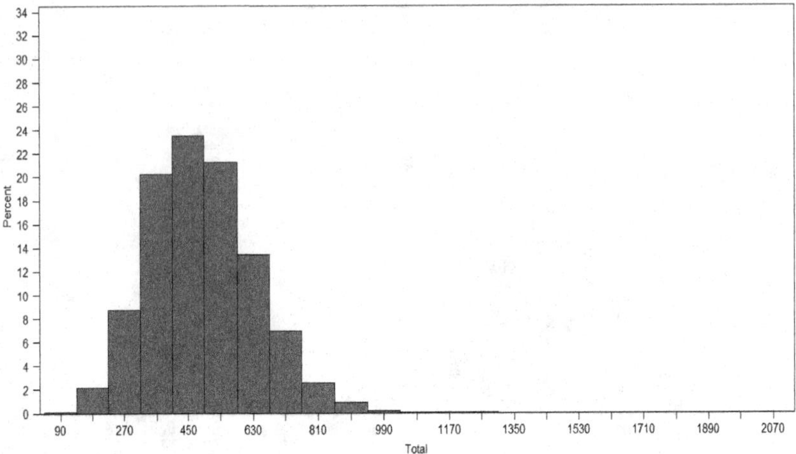

Figure 7.22 *Histogram of total operational loss in our example of publicly available data: Weibull model with underreporting.*

```
*...(continues from the previous page)...;

/*Boundary correction*/
  start grkorr(x,b);
    upperlim=min(1,(1-x)/b); lowerlim=max(-1,-x/b);
    rr=kint(upperlim)-kint(lowerlim);
    return (rr);
  finish;
/*Kernel density with boundary correction for
  the transformed data*/
  start fhatz(z,v,nr,b);
    /*z is a scalar, v is a vector*/
    rr=sum(K((z*j(nr,1,1)-v)/b))/(nr*b*grkorr(z,b));
    return(rr);
  finish;
/*Transformed Kernel density estimation*/
  start fhatx(x,v,nr,b,a,s);
    Tcapx=Tcap(x,a,s);
    density=pdf('WEIBULL',x,a,s);
    rr=fhatz(Tcapx,v,nr,b)*density;
    return(rr);
  finish;

  *...(continues on the next page)...;
```

```
*...(continues from the previous page)...;

/*Density estimation of reported claims*/
f_R=j(nx,ncat,0);
sx=j(R,ncat,0);
n0=0;
do k=1 to ncat;
  /*Parameters of Weibull distribution*/
  a=Par_weib[1,k];
  s=Par_weib[2,k];
  nl=n0+n[k];
  xk=xall[n0+1:nl];
  z=Tcap(xk,a,s);
  sigma=sqrt((sum((z-(sum(z)/n[k])#j
             (n[k],1,1))##2))/n[k]);
  b=2.34*sigma*(n[k]**(-1/5));
  n0=nl;
  do m=1 to nx;
    f_R[m,k]=fhatx(x[m],z,n[k],b,a,s);
  end;
  do i=1 to R;
    if lambda_ri[i,k]>0 then do;
      u=uniform(j(lambda_ri[i,k],1,100+i));
      /*Calculate simulated internal losses*/
      gx=j(lambda_ri[i,k],1,0);
      do j=1 to lambda_ri[i,k];
        u2=f_R[1,k]*precis;
        do l=2 to ngrid until(u2>u[j]);
          u2=u2+f_R[l,k]*precis;
          gx[j]=x[l];
      end; end;
      sx[i,k]=sum(gx);
end; end; end;
totsx=sx[,+];
res=sx||totsx;

/*Data base with total cost*/
create oper.WE_KS
       from res [colname={'s_1' 's_2' 'Total'}];
append from res;
close oper.WE_KS;

*...(continues with the model WE.KS.UR)...;
```

```
axis1 order=(0 to 44 by 2);

proc univariate data=oper.WE_KS noprint;
 var s_1;
 histogram /vaxis=axis1
           midpoints=30 90 150 210 270 330 390 450
                     510 570 630 690 750 810 870
                     930 990 1050 1110 1170 1230
                     1290 1350 1410 1470 1530 1590
                     1650 1710 1770;
run;

axis1 order=(0 to 30 by 2);

proc univariate data=oper.WE_KS noprint;
 var s_2;
 histogram/vaxis=axis1
          midpoints=20 60 100 140 180 220 260 300 340
                    380 420 460 500 540 580 620 660
                    700 740 780 820 860 900 940 980
                    1020 1060 1100 1140;
run;

axis1 order=(0 to 34 by 2);

proc univariate data=oper.WE_KS noprint;
 var Total;
 histogram/vaxis=axis1
          midpoints=90 180 270 360 450 540 630 720 810
                    900 990 1080 1170 1260 1350 1440
                    1530 1620 1710 1800 1890 1980 2070;
run;
```

In Figures 7.23, 7.24, and 7.25 we show histograms of operational loss for each risk category and for the sum, assuming that individual loss are generated from a transformed kernel estimation using Weibull distribution as transformation and without underreporting. Histograms are now smoother.

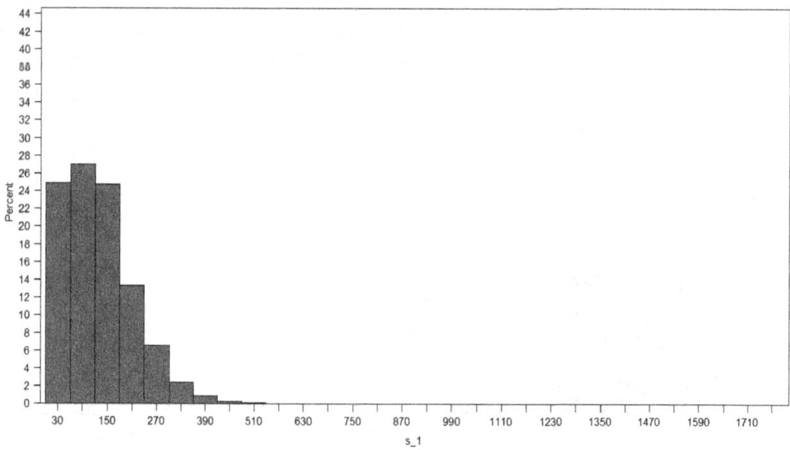

Figure 7.23 *Histogram of operational loss in our example of publicly available data for risk category 1: Transformed kernel estimation using Weibull transformation model without underreporting.*

```
*...(continues from the model WE.KS)...;

 /*MODEL WE.KS.UR*/
  /*Density of occurred claims*/
    f_O=j(nx,ncat,0);
    precis=0.01;
    do k=1 to ncat;
      /*Numerator*/
      num=f_R[,k]/und[,k];
      /*Denominator*/
      den=sum((f_R[,k]/und[,k])#precis);
      f_O[,k]=num/den;
    end;
  /*Probability of reporting a claim*/
    P_g=j(1,ncat,0);
    do k=1 to ncat;
      P_g[k]=sum((f_O[,k]#und[,k])#precis);
    end;
    print P_g;
  *...(continues on the next page)...;
```

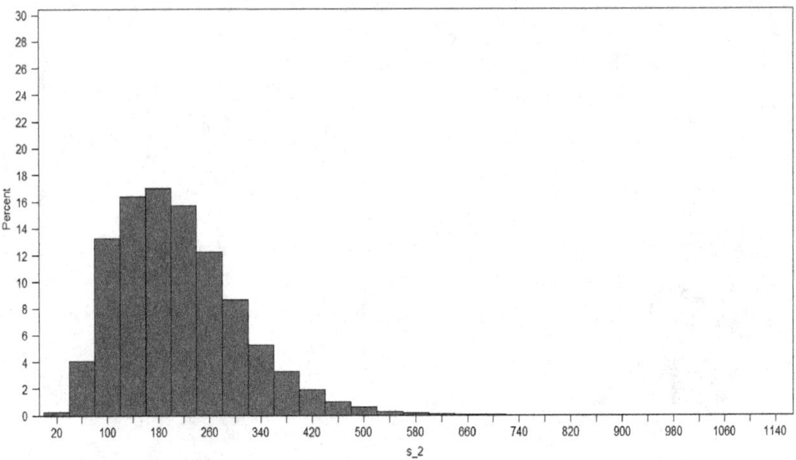

Figure 7.24 *Histogram of operational loss in our example of publicly available data for risk category 2: Transformed kernel estimation using Weibull transformation model without underreporting.*

```
  *...(continues from the previous page)...;

/*Number of occurred claims*/
  /*Frequency in each reported claims risk category*/
  lambda_i={10,20};
  /*Frequency in each occurred claims risk category*/
  lambda_O=lambda_i/P_g‘;

/*Results Table 2 file We.UR*/
  print lambda_O;
  lambda_oi=j(R,ncat,0);
  do k=1 to ncat;
    lambda_oi[,k]=quantile('POISSON',ran,lambda_o[k]);
  end;
  /*Total cost results with underreporting*/
  sx=j(R,ncat,0);
  do k=1 to ncat;
    do i=1 to R;
      if lambda_oi[i,k]>0 then do;
        u=uniform(j(lambda_oi[i,k],1,100+i));
        /*Calculate simulated internal losses*/
        gx=j(lambda_oi[i,k],1,0);

*...(continues on the next page)...;
```

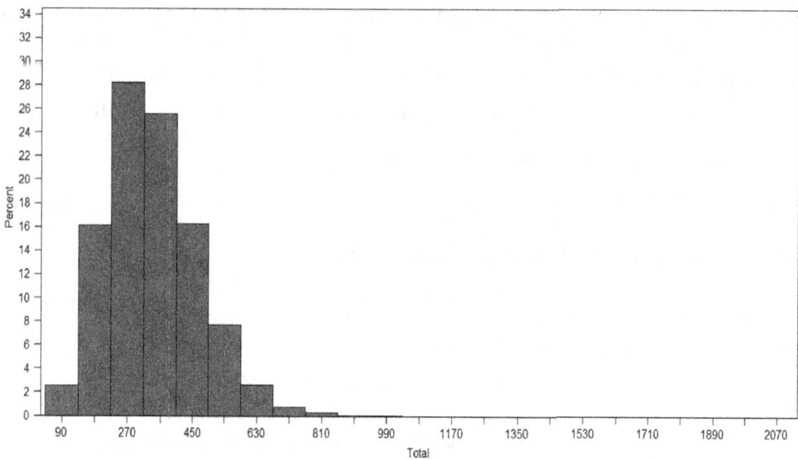

Figure 7.25 *Histogram of total operational loss in our example of publicly available data: Transformed kernel estimation using Weibull transformation model without underreporting.*

```
    *...(continues from the previous page)...;

          do j=1 to lambda_oi[i,k];
            u2=f_O[1,k]*precis;
            do l=2 to nx until(u2>u[j]);
              u2=u2+f_O[l,k]*precis;
              gx[j]=x[l];
            end;
          end;
          sx[i,k]=sum(gx);
        end;
      end;
    end;
    totsx=sx[,+];
    res=sx||totsx;

    /*Data set with total cost*/
    create oper.WE_KS_UR
           from res [colname={'s_1' 's_2' 'Total'}];
    append from res;
    close oper.WE_KS_UR;
quit;
```

Finally, in Figures 7.26, 7.27, and 7.28 we show histograms of operational loss for each risk category and for their sum, assuming that individual loss are generated with a transformed kernel estimation using Weibull distribution as transformation and with underreporting. This model combines the correction for the distributional assumption and the correction for underreporting.

```
axis1 order=(0 to 44 by 2);

proc univariate data=oper.WE_KS_UR noprint;
 var s_1;
 histogram /vaxis=axis1
           midpoints=30 90 150 210 270 330 390 450
                     510 570 630 690 750 810 870
                     930 990 1050 1110 1170 1230
                     1290 1350 1410 1470 1530 1590
                     1650 1710 1770;
 run;

axis1 order=(0 to 30 by 2);

proc univariate data=oper.WE_KS_UR noprint;
 var s_2;
 histogram/vaxis=axis1
          midpoints=20 60 100 140 180 220 260 300 340
                    380 420 460 500 540 580 620 660
                    700 740 780 820 860 900 940 980
                    1020 1060 1100 1140;

run;

axis1 order=(0 to 34 by 2);

proc univariate data=oper.WE_KS_UR noprint;
 var Total;
 histogram/vaxis=axis1
          midpoints=90 180 270 360 450 540 630 720 810
                    900 990 1080 1170 1260 1350 1440
                    1530 1620 1710 1800 1890 1980 2070;
run;
```

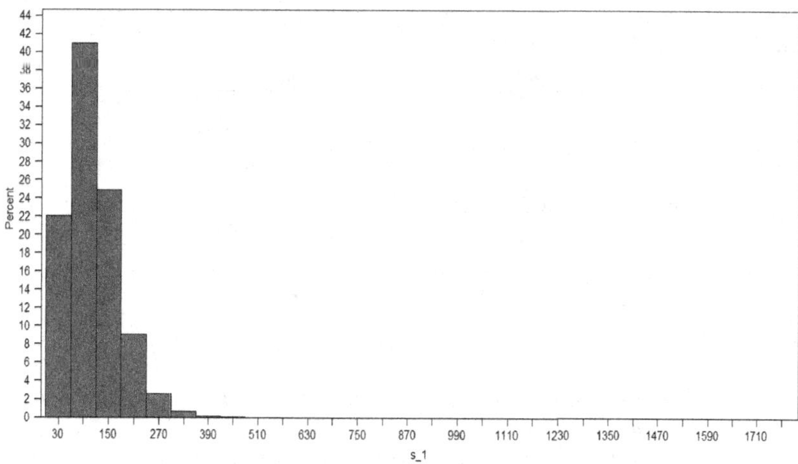

Figure 7.26 *Histogram of operational loss in our example of publicly available data for risk category 1: Transformed kernel estimation using Weibull transformation model with underreporting.*

The results that correspond to the methods in Chapter 5 are shown in Table 7.29. In the table, n is the number of losses, Par_Weib are the estimated parameters, and P_g in row four corresponds to the probability of reported claims with a Weibull parametric approach. In row five we present lambda_i, that is, the number of reported claims, hereafter the result denoted by lambda_0, the number of occurred claims (Weibull parametric approach). In row six, P_g, is the probability of reported claims with a transformed kernel estimation with a cumulative distribution function (cdf) Weibull transformed approach. Finally, the last row, lambda_0, is the number of occurred claim claims with a transformed kernel estimation with a cdf Weibull transformed approach.

Using similar tools as the ones described in the previous section, the VaR and the TVaR measures are found in Tables 7.30 to 7.35. The first two tables present the results with the Weibull approach (Model WE), while the final two tables show the Weibull parametric approach with underreporting (Model WE_UR).

Table 7.29 *Results for the methods described in Chapter 5 and our example of publicly available data risk categories 1 and 2*

Parameter	Data no. 1	Data no. 2
n	1000	400
Par_weib	(0.51607,11.59642)	(0.61703,10.48704)
P_g	0.5953199	0.6275121
lambda_i	10	20
lambda_0	16.797692	31.871895
P_g	0.6998422	0.7077762
lambda_0	14.288935	28.257518

Table 7.30 *General statistics for model WE in Chapter 5 and our example of publicly available data risk categories 1 and 2*

Data	Mean	Median	Sd
S_1	220.545	181.782	26949.977
S_2	307.243	288.461	18305.386
S_total	527.788	495.663	46068.263

Table 7.31 *Quantile statistics for the model WE in Chapter 5 and our example of publicly available data risk categories 1 and 2*

	VaR95	VaR99	VaR999
S_1	536.783	782.526	1152.160
S_2	557.629	702.550	881.642
S_total	920.870	1185.562	1549.664

Table 7.32 *Tail-Value-at-Risk for the model WE in Chapter 5 and our example of publicly available data risk categories 1 and 2*

	TVaR95	TVaR99	TVaR999
S_1	692.117	963.155	1477.022
S_2	648.766	794.954	1058.025
S_total	1088.675	1360.731	1907.596

Table 7.33 *General statistics for model WE_UR in Chapter 5 and our example of publicly available data risk categories 1 and 2*

Data	Mean	Median	Sd
S_1	241.086	201.410	28044.014
S_2	342.979	324.890	18561.511
S_total	584.065	548.115	53574.367

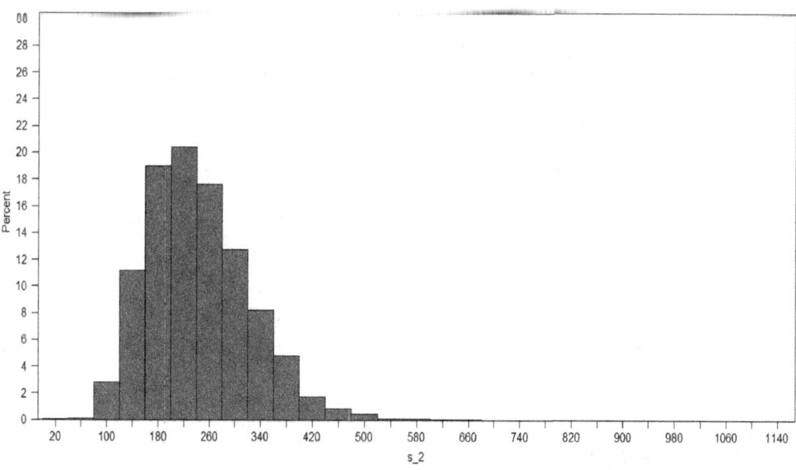

Figure 7.27 *Histogram of operational loss in our example of publicly available data for risk category 2; Transformed kernel estimation using Weibull transformation model with underreporting.*

Table 7.34 *Quantile statistics for the model WE_UR in Chapter 5 and our example of publicly available data risk categories 1 and 2*

	VaR95	VaR99	VaR999
S_1	554.23	802.29	1256.02
S_2	554.23	802.29	1256.02
S_total	998.5	1274.39	1626.42

Table 7.35 *Tail-Value-at-Risk for the model WE_UR in Chapter 5 our example of publicly available data risk categories 1 and 2*

	TVaR95	TVaR99	TVaR999
S_1	719.862	1000.452	1685.421
S_2	719.862	1000.452	1685.421
S_total	1178.337	1470.236	2134.564

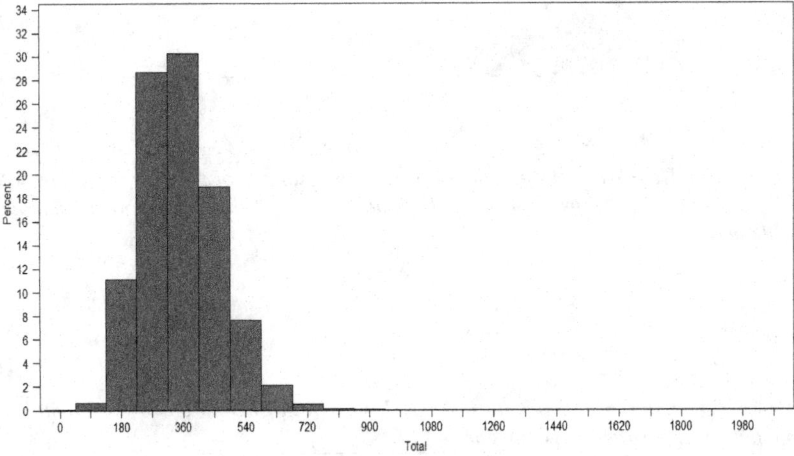

Figure 7.28 *Histogram of total operational loss in our example of publicly available data: Transformed kernel estimation using Weibull transformation model with underreporting.*

7.6 Programming in R

7.6.1 Basic Functions

The procedures described in the previous sections are also very easily programmed in R. Here are a few examples of the core function parts that can be used for immediate implementation in programming language R.

```
# FUNCTIONS:
# SORT a matrix with respect to COLUMN scol

SortMatrix = function(Matrix,scol)
{
NN = nrow(Matrix)
NewMatrix = matrix(nrow=NN,ncol=ncol(Matrix))
r = order(Matrix[,scol])
for (i in 1:NN)
{ NewMatrix[i,] = Matrix[r[i],] }
return(NewMatrix) }

# GENERALIZED CHAMPERNOWNE DISTRIBUTION
pchamp = function(x,par)
{
a=par[1]
M=par[2]
c=par[3]
((x+c)^a-c^a)/((x+c)^a+(M+c)^a-2*c^a) }

dchamp = function(x,par)
{
a=par[1]
M=par[2]
c=par[3]
(a*(x+c)^(a-1)*((M+c)^a-c^a))/((x+c)^a+(M+c)^a-2*c^a)^2 }

qchamp = function(y,par)
{
a=par[1]
M=par[2]
c=par[3]
((y*(M+c)^a+(1-2*y)*c^a)/(1-y))^(1/a)-c }

champLike = function(data,par)
{
a=par[1]
M=par[2]
c=par[3]
N = length(data)
N*log(a)+N*log((M+c)^a-c^a)+(a-1)*sum(log(data+c))
```

```
+    -2*sum(log((data+c)^a+(M+c)^a-2*c^a)) }

champPar = function(data)
{
N = length(data)
par = matrix(0,nrow=3,ncol=1)
L = 0
avalg = 0
Lvalg = 0
cvalg = seq(0,2,0.5)
par[2] = median(data)
seqp = seq(1,3,1)
seqj = seq(1,21,1)
# Finder optimalt a for hver vrdi af c
for (i in seq(1,length(cvalg),1))
{
par[3]=cvalg[i]*par[2]
astart=0.1
aslut=21
for (praecision in seqp)
{
for (j in seqj)
{
par[1]=astart+(j-1)*(aslut-astart)/20
L[j]=champLike(data,par) }
jmax=which.max(L)
astart=max(0.0001,astart+(jmax-2)
+                    *(aslut-astart)/20)
aslut=astart+jmax*(aslut-astart)/20 }
avalg[i]=astart+(jmax-1)*(aslut-astart)/20
Lvalg[i]=L[jmax] }
iopt=which.max(Lvalg)
par[1]=avalg[iopt]
par[3]=cvalg[iopt]*par[2]
return(par) }

# FUNCTIONS FOR KERNEL ESTIMATION
# The normal scale bandwidth selection with Epanechnikov kernel
BdNormEpan = function(data)
{
((40*pi^(1/2))/length(data))^(1/5)*sd(data) }

# Kernel function
# Epanechnikov kernel function in one dimension
Epan = function(u)
{
0.75*(1-u^2)*(abs(u)<=1) }
```

```
# Boundary correction for data between 0 and 1
Grkorr01 = function(x,b)
{
# Linear
upperlim=pmin(1,x/b);
lowerlim=pmax(-1,(x-1)/b);
KintUpper = 0.25*(3-upperlim^2)*upperlim;
KintLower = 0.25*(3-lowerlim^2)*lowerlim;
kor = KintUpper-KintLower;
return(kor) }

Grkorr02 = function(x,b)
{
# Non linear
upperlim=pmin(1,x/b);
lowerlim=pmax(-1,(x-1)/b);
KintUpper = 0.75*( 0.5-0.25*((upperlim)^2) )*(upperlim)^2;
KintLower = 0.75*( 0.5-0.25*((lowerlim)^2) )*(lowerlim)^2;
kor = KintUpper-KintLower;
return(kor) }

Grkorr03 = function(x,b)
{
# Non linear
upperlim=pmin(1,x/b);
lowerlim=pmax(-1,(x-1)/b);
KintUpper = 0.25*( 1-0.6*((upperlim)^2) )*((upperlim)^3);
KintLower = 0.25*( 1-0.6*((lowerlim)^2) )*((lowerlim)^3);
kor = KintUpper-KintLower;
return(kor) }

Grkorr04 = function(x,b)
{
# Non linear
upperlim=pmin(1,x/b);
lowerlim=pmax(-1,(x-1)/b);
KintUpper = 0.75*( 0.25-(1/6)*((upperlim)^2) )*((upperlim)^4);
KintLower = 0.75*( 0.25-(1/6)*((lowerlim)^2) )*((lowerlim)^4);
kor = KintUpper-KintLower;
return(kor) }

TrKerDen = function(x, data, distr, par, band, kernel){
# Input: x : vector between [0,Inf)
# data    : Original Data
# distr   : Distribution "champ", "lnorm" or "weibull"
# par     : Estimated Parameters of the 'distr'
```

```
# band    : bandwidth
# kernel : Local Constant ("LC") or Local Linear ("LL") kernel
# Output: Den : Density estimator
# Quan    : Quantile estimator
# Control : Integration value

dchamp=function(x,a=par[1],M=par[2],c=par[3]){
        (a*(x+c)^(a-1)*((M+c)^a-c^a))/((x+c)^a+(M+c)^a-2*c^a)^2 }
if(distr == "champ"){
Trans = function(x,par) pchamp(x,par)
dataTrans = pchamp(data,par)
Transinv =  qchamp
trans = function(x,par) dchamp(x,a=par[1],M=par[2],c=par[3])  }
if(distr == "lnorm"){
Trans = function(x,par) plnorm(x, meanlog = par[1],
          + sdlog = par[2], lower.tail = TRUE, log.p = FALSE)
dataTrans = plnorm(data,meanlog = par[1], sdlog = par[2],
          + lower.tail = TRUE, log.p = FALSE)
Transinv =  qlnorm(p, meanlog = par[1], sdlog = par[2],
          + lower.tail = TRUE, log.p = FALSE)
trans = function(x,par) dlnorm(x,meanlog = par[1],
          + sdlog = par[2], log = FALSE) }
if(distr == "weibull"){
Trans = function(x,par) pweibull(x,shape = par[1],
          + scale = par[2], lower.tail = TRUE, log.p = FALSE)
dataTrans = pweibull(data,  shape = par[1], scale = par[2],
          + lower.tail = TRUE, log.p = FALSE)
Transinv =  qweibull(p,shape = par[1], scale = par[2],
          + lower.tail = TRUE, log.p = FALSE)
trans = function(x,par) dweibull(x,shape = par[1],
          + scale = par[2], log = FALSE) }
xTrans = c(Trans(x,par),1)
    if(kernel == "LC") TransEst = KerDen(xTrans,dataTrans, band)
    if(kernel == "LL") TransEst = KerDenLL(xTrans,dataTrans, band)
Control = TransEst$Control
#Transformation in original axis
Den = function(x)
{
TransEst$Den(Trans(x,par)) * trans(x,par) }

# Quantile function
Quan = function(q)
{
Transinv(TransEst$Quan(q), par) }

# return(Den, Quan, Control)
return(Den, Control) }
```

```
KerDen = function(x, data, band)
{
#       Inputs:     data    : Transformed Data
#                   x       : vector between [0,1]
#                   band    : bandwidth
#
#       Outputs:    Den     : Density estimator
#                   Quan    : Quantile estimator
#                   Control : Integration value

Grkorr =Grkorr01

# Endimensional tthed
b = band
xM = rep(1,times=length(data)) %o% x
dataM = data %o% rep(1,times=length(x))
arg = Epan((xM-dataM)/b)
Density = colSums(arg)
Density = Density / (Grkorr(x,b)*length(data)*b)

# Change made for OR-II article: changed splinefun to approxfun
Densityapprox = approxfun(x, Density)
Control = integrate(Densityapprox,0,1)[[1]]
# Density function
Den = function(x)
{
Densityapprox(x)/Control }

# Quantile function
Integrant = function(m,q)
{
integrate(Den,0,m)[[1]]-q }

Quan = function(q)
{
uniroot(function (m) Integrant(m,q), c(0,1))$root }

return(Den, Quan, Control)
}
```

After seeing the basic functions, we show how to produce the plots in Chapter 3.

```
# Plotting the Transformation (Champernowne) Local Constant
# Kernel Density Estimator Process in [0,1]

# Transform Data
internal.data.tr<-pchamp(internal.data,par.internal)
```

```
external.data.tr<-pchamp(external.data,par.external)

# KDE
KerDen.internal<-KerDen(u,internal.data.tr,
                        band(internal.data.tr))$Den
KerDen.external<-KerDen(u,external.data.tr,
                        band(external.data.tr))$Den

# Figure - transformed Losses
par(mfrow=c(1,2))
hist(internal.data.tr,nclass=60,freq=0,xlab="Transformed Data",
+    main="Transformed Internal Data")
hist(external.data.tr,nclass=60,freq=0,xlab="Transformed Data",
+    main="Transformed External Data")

# Figure - including the KDE on the histogram
par(mfrow=c(1,2))
hist(internal.data.tr,nclass=60,freq=0,xlab="Transformed Data",
+    main="Estimated KDE on Transformed Internal Data ")
lines(u,KerDen.internal(u))
hist(external.data.tr,nclass=60,freq=0,xlab="Transformed Data",
+    main="Estimated KDE on Transformed External Data")
lines(u,KerDen.external(u))

# Plotting the Transformation (Champernowne) Local Constant Kernel
# Density Estimator [0,Inf)

KerDen.internal.tr<-TrKerDen(x,internal.data,"champ",par.internal,
+      band(internal.data.tr),"LC")$Den
KerDen.external.tr<-TrKerDen(x,external.data,"champ",par.external,
+      band(external.data.tr),"LC")$Den

par(mfrow=c(2,2))
plot(x,KerDen.internal.tr(x),type="l",col=1,xlab="Size",
+    xlim=c(0,5),ylim=c(0,0.9),ylab="Probability",
     main='Estimated KDE for Internal Data',
     cex.main=0.9,col.main=1)
points(internal.data,x0.internal,col=1)
lines(x,Den.parametric.internal,lty=2)
#legend("topright",  c("KDE","GCD") ,cex=0.8,lty=c(1,2),bty="n",
+       col=c(1,1))

plot(x,KerDen.internal.tr(x),type="l",col=1,xlab="Size",
     xlim=c(5,35),ylim=c(0,0.01),ylab="Probability",
     main='Estimated KDE for Internal Data',cex.main=0.9,
     col.main=1)
points(internal.data,x0.internal,col=1)
```

```
lines(x,Den.parametric.internal,lty=2)
#legend("topright",  c("KDE","GCD") ,cex=0.8,lty=c(1,2),bty="n",
+        col=c(1,1))

plot(x,KerDen.external.tr(x),type="l",col=1,xlab="Size",
     xlim=c(0,5),ylim=c(0,0.9),ylab="Probability",
     main='Estimated KDE for External Data',cex.main=0.9,
     col.main=1)
points(external.data,y0.external,col=1)
lines(x,Den.parametric.external,lty=2)
#legend("topright",  c("KDE","GCD") ,cex=0.8,lty=c(1,2),
+        bty="n",col=c(1,1))

plot(x,KerDen.external.tr(x),type="l",col=1,xlab="Size",
     xlim=c(5,35),ylim=c(0,0.01),ylab="Probability",
     main='Estimated KDE for External Data',cex.main=0.9,
     col.main=1)
points(external.data,y0.external,col=1)
lines(x,Den.parametric.external,lty=2)
#legend("topright",  c("KDE","GCD") ,cex=0.8,lty=c(1,2),
+        bty="n",col=c(1,1))
```

The procedures presented in Chapter 6 can also easily be implemented with the basic function program that have been provided in this chapter.

7.6.2 *Figures in Chapters 2 and 3*

In this section we present the R programs that are used to obtain the plots in Chapters 2 and 3. The results are now illustrated with the data sets that have been presented here, so the plots correspond to the guided example data, and they should not be confused with the ones shown in Chapters 2 and 3.

Figures in Chapter 2 can be reproduced with the program shown below:

```
internal.data<-read.table("int_sim_tabla2_1.txt",header=TRUE)
external.data<-read.table("ext_sim_tabla2_1.txt",header=TRUE)

### Plotting the Data used in the book (Figure 2.1)
x.internal<-c(1:length(internal.data$y))
y.external<-c(1:length(external.data$y))

par(mfrow=c(1,2))
plot(x.internal,internal.data$y,type="p",col=1,xlab="",ylab="Size",
     main='Internal Losses',cex.main=0.9,col.main=1)
plot(y.external,external.data$y,type="p",col=1,xlab="",ylab="Size",
     main='External Losses',cex.main=0.9,col.main=1)


library(MASS)
```

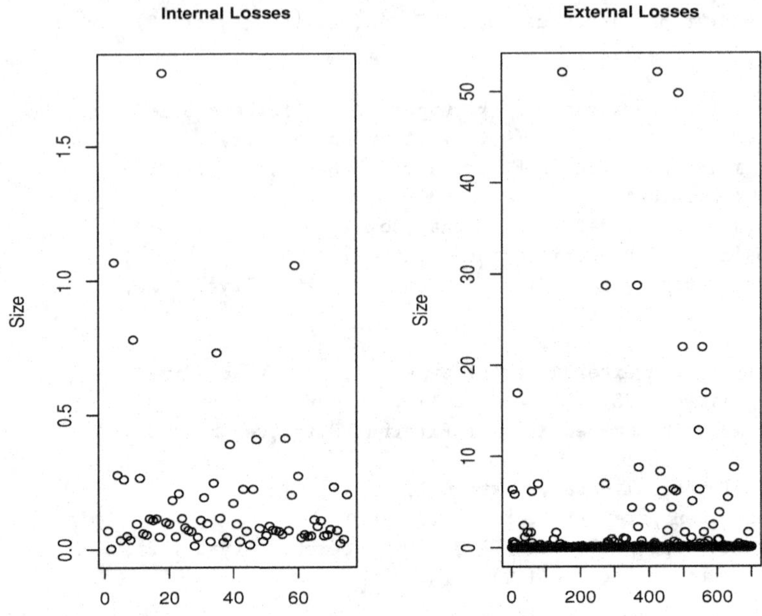

Figure 7.29 *Plotted internal and external operational risk example data originating from the event risk category "Business disruption and system failures".*

```
### Plotting the Estimated Logisitic Distribution (Figure 2.2)

x=seq(0,100,length=2000)
xx=seq(0,5,length=2000)
u=seq(0,1,length=2000)

x0.internal=rep(0,length(internal.data$y))
y0.external=rep(0,length(external.data$y))

int.data<-as.vector(internal.data$y)
ext.data<-as.vector(external.data$y)

par.internal.L<-fitdistr(internal.data$y,"logistic")
par.external.L<-fitdistr(external.data$y,"logistic")

Den.parametric.internal.L<-dlogis(x,par.internal.L[[1]][1],
                                  par.internal.L[[1]][2])
Den.parametric.external.L<-dlogis(x,par.external.L[[1]][1],
                                  par.external.L[[1]][2])
```

```
par(mfrow=c(2,2))
plot(x,Den.parametric.internal.L,type="l",col=1,xlab="Size",
     xlim=c(0,5),ylim=c(0,0.9),ylab="Probability",
     main='Internal Logistic Distribution',cex.main=0.9,
     col.main=1)
points(internal.data$y,x0.internal,col=1)

plot(x,Den.parametric.internal.L,type="l",col=1,xlab="Size",
     xlim=c(5,35),ylim=c(0,0.01),ylab="Probability",
     main='Internal Logistic Distribution',cex.main=0.9,
     col.main=1)
points(internal.data$y,x0.internal,col=1)

plot(x,Den.parametric.external.L,type="l",col=1,xlab="Size",
     xlim=c(0,5),ylim=c(0,0.9),ylab="Probability",
     main='External Logistic Distribution',cex.main=0.9,
     col.main=1)
points(external.data$y,y0.external,col=1)

plot(x,Den.parametric.external.L,type="l",col=1,xlab="Size",
     xlim=c(5,35),ylim=c(0,0.01),ylab="Probability",
     main='External Logistic Distribution',cex.main=0.9,
     col.main=1)
points(external.data$y,y0.external,col=1)


### Plotting the Estimated Weibull Distribution (Figure 2.3)

par.internal.W<-fitdistr(internal.data$y,"weibull")
par.external.W<-fitdistr(external.data$y,"weibull")

Den.parametric.internal.W<-dweibull(x,par.internal.W[[1]][1],
                               par.internal.W[[1]][2])
Den.parametric.external.W<-dweibull(x,par.external.W[[1]][1],
                               par.external.W[[1]][2])

par(mfrow=c(2,2))
plot(x,Den.parametric.internal.W,type="l",col=1,xlab="Size",
     xlim=c(0,5),ylim=c(0,0.9),ylab="Probability",
     main='Internal Weibull Distribution',cex.main=0.9,
     col.main=1)
points(internal.data$y,x0.internal,col=1)

plot(x,Den.parametric.internal.W,type="l",col=1,xlab="Size",
     xlim=c(5,35),ylim=c(0,0.01),ylab="Probability",
     main='Internal Weibull Distribution',cex.main=0.9,
     col.main=1)
```

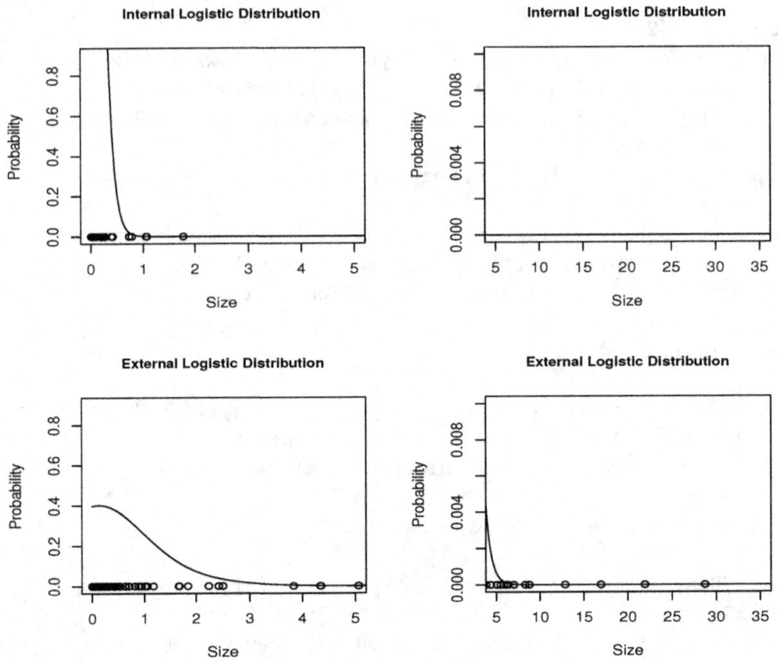

Figure 7.30 *The estimated density for internal example losses (above) and external example losses (below) operational risk data using the logistic distribution.*

```
points(internal.data$y,x0.internal,col=1)

plot(x,Den.parametric.external.W,type="l",col=1,xlab="Size",
     xlim=c(0,5),ylim=c(0,0.9),ylab="Probability",
     main='External Weibull Distribution',cex.main=0.9,
     col.main=1)
points(external.data$y,y0.external,col=1)

plot(x,Den.parametric.external.W,type="l",col=1,xlab="Size",
     xlim=c(5,35),ylim=c(0,0.01),ylab="Probability",
     main='External Weibull Distribution',cex.main=0.9,
     col.main=1)
points(external.data$y,y0.external,col=1)
```

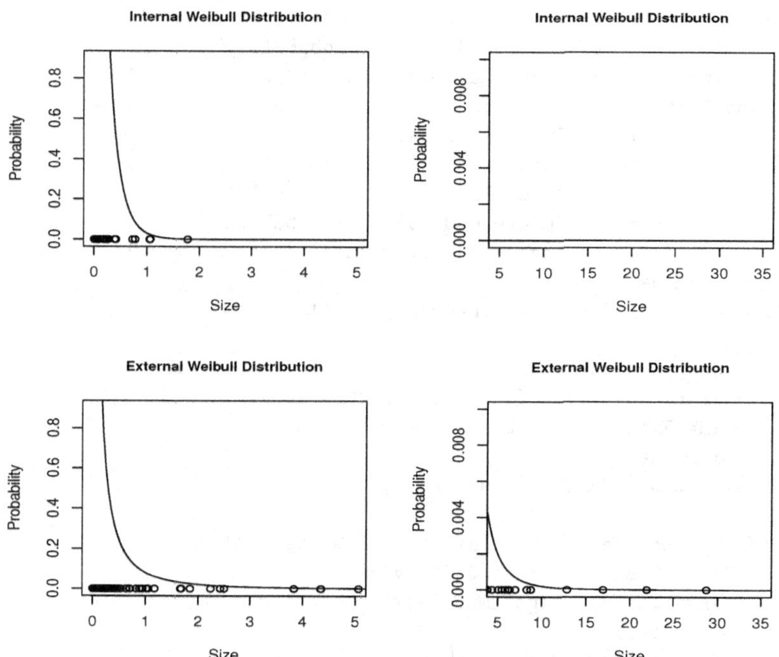

Figure 7.31 *The estimated density for internal example losses (above) and external example losses (below) operational risk data using the Weibull distribution.*

```
### Plotting the Estimated lognormal Distribution (Figure 2.4)

par.internal.LN<-fitdistr(internal.data$y,"log-normal")
par.external.LN<-fitdistr(external.data$y,"log-normal")

Den.parametric.internal.LN<-dlnorm(x,par.internal.LN[[1]][1],
                            par.internal.LN[[1]][2])
Den.parametric.external.LN<-dlnorm(x,par.external.LN[[1]][1],
                            par.external.LN[[1]][2])

par(mfrow=c(2,2))
plot(x,Den.parametric.internal.LN,type="l",col=1,xlab="Size",
     xlim=c(0,5),ylim=c(0,0.9),ylab="Probability",
     main='Internal Lognormal Distribution',cex.main=0.9,
     col.main=1)
points(internal.data$y,x0.internal,col=1)

plot(x,Den.parametric.internal.LN,type="l",col=1,xlab="Size",
     xlim=c(5,35),ylim=c(0,0.01),ylab="Probability",
     main='Internal Lognormal Distribution',cex.main=0.9,
     col.main=1)
points(internal.data$y,x0.internal,col=1)

plot(x,Den.parametric.external.LN,type="l",col=1,xlab="Size",
     xlim=c(0,5),ylim=c(0,0.9),ylab="Probability",
     main='External Lognormal Distribution',cex.main=0.9,
     col.main=1)
points(external.data$y,y0.external,col=1)

plot(x,Den.parametric.external.LN,type="l",col=1,xlab="Size",
     xlim=c(5,35),ylim=c(0,0.01),ylab="Probability",
     main='External Lognormal Distribution',cex.main=0.9,
     col.main=1)
points(external.data$y,y0.external,col=1)



### Plotting Champernowne Distribution (Figure 2.5)
##Function definition
pchamp = function(x,par)
{
a=par[1]
M=par[2]
c=par[3]
((x+c)^a-c^a)/((x+c)^a+(M+c)^a-2*c^a)
}

dchamp = function(x,par)
```

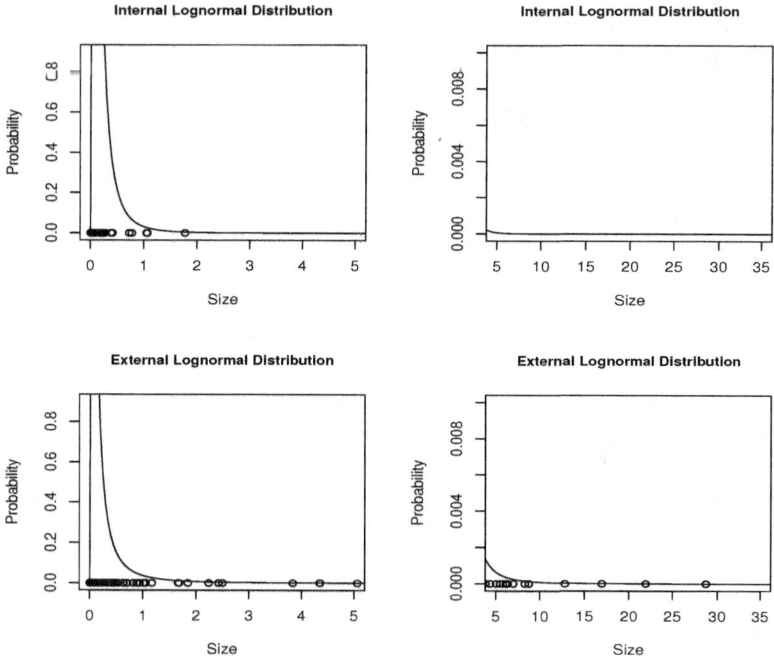

Figure 7.32 *The estimated density for internal example losses (above) and external example losses (below) operational risk data using the lognormal distribution.*

```
{
a=par[1]
M=par[2]
c=par[3]
(a*(x+c)^(a-1)*((M+c)^a-c^a))/((x+c)^a+(M+c)^a-2*c^a)^2
}


qchamp = function(y,par)
{
a=par[1]
M=par[2]
c=par[3]
((y*(M+c)^a+(1-2*y)*c^a)/(1-y))^(1/a)-c
}

champLike = function(data,par)
{
a=par[1]
```

```
M=par[2]
c=par[3]
N = length(data)
N*log(a)+N*log((M+c)^a-c^a)+(a-1)*sum(log(data+c))-
    2*sum(log((data+c)^a+(M+c)^a-2*c^a))
}

champPar = function(data)
{
N = length(data)
par = matrix(0,nrow=3,ncol=1)
L = 0
avalg = 0
Lvalg = 0
cvalg = seq(0,2,0.5)
par[2] = median(data)
seqp = seq(1,3,1)
seqj = seq(1,21,1)
# Finder optimalt a for hver vrdi af c
for (i in seq(1,length(cvalg),1))
{
par[3]=cvalg[i]*par[2]
astart=0.1
aslut=21
for (praecision in seqp)
{
for (j in seqj)
{
par[1]=astart+(j-1)*(aslut-astart)/20
L[j]=champLike(data,par)
}
jmax=which.max(L)
astart=max(0.0001,astart+(jmax-2)*(aslut-astart)/20)
aslut=astart+jmax*(aslut-astart)/20
}
avalg[i]=astart+(jmax-1)*(aslut-astart)/20
Lvalg[i]=L[jmax]
}
iopt=which.max(Lvalg)
par[1]=avalg[iopt]
par[3]=cvalg[iopt]*par[2]
return(par)
}

x1=seq(0,50,length=1000)
par<-c(0.5,2,0)
cdf11d<-pchamp(x1,par)
```

```
pdf11d<-dchamp(x1,par)
par<-c(0.5,2,2)
cdf11s<-pchamp(x1,par)
pdf11s<-dchamp(x1,par)
par<-c(1,2,0)
cdf21d<-pchamp(x1,par)
pdf21d<-dchamp(x1,par)
par<-c(1,2,2)
cdf21s<-pchamp(x1,par)
pdf21s<-dchamp(x1,par)
par<-c(2,2,0)
cdf31d<-pchamp(x1,par)
pdf31d<-dchamp(x1,par)
par<-c(2,2,2)
cdf31s<-pchamp(x1,par)
pdf31s<-dchamp(x1,par)

par(mfrow=c(2,3))

plot(x1,cdf11s,type="l",col=1,xlab="Size",xlim=c(0,50),
     ylim=c(0,1),ylab="",main='a) Cdf, alpha=0.5, M=2',
     cex.main=0.9,col.main=1)
lines(x1,cdf11d,type="l",lty=2,col=1)
plot(x1,pdf11s,type="l",col=1,xlab="Size",xlim=c(0,50),
     ylim=c(0,0.5),ylab="",main='b) Pdf, alpha=0.5, M=2',
     cex.main=0.9,col.main=1)
lines(x1,pdf11d,type="l",lty=2,col=1)
plot(x1,cdf21s,type="l",col=1,xlab="Size",xlim=c(0,50),
     ylim=c(0,1),ylab="",main='c) Cdf, alpha=1.0, M=2',
     cex.main=0.9,col.main=1)
lines(x1,cdf21d,type="l",lty=2,col=1)
plot(x1,pdf21s,type="l",col=1,xlab="Size",xlim=c(0,50),
     ylim=c(0,0.5),ylab="",main='d) Pdf, alpha=1.0, M=2',
     cex.main=0.9,col.main=1)
lines(x1,pdf21d,type="l",lty=2,col=1)
plot(x1,cdf31s,type="l",col=1,xlab="Size",xlim=c(0,50),
     ylim=c(0,1),ylab="",main='e) Cdf, alpha=2.0, M=2',
     cex.main=0.9,col.main=1)
lines(x1,cdf31d,type="l",lty=2,col=1)
plot(x1,pdf31s,type="l",col=1,xlab="Size",xlim=c(0,50),
     ylim=c(0,0.5),ylab="",main='f) Pdf, alpha=2.0, M=2',
     cex.main=0.9,col.main=1)
lines(x1,pdf31d,type="l",lty=2,col=1)


### Plotting the Estimated Champernowne Distribution (Figure 2.6)

x=seq(0,100,length=2000)
```

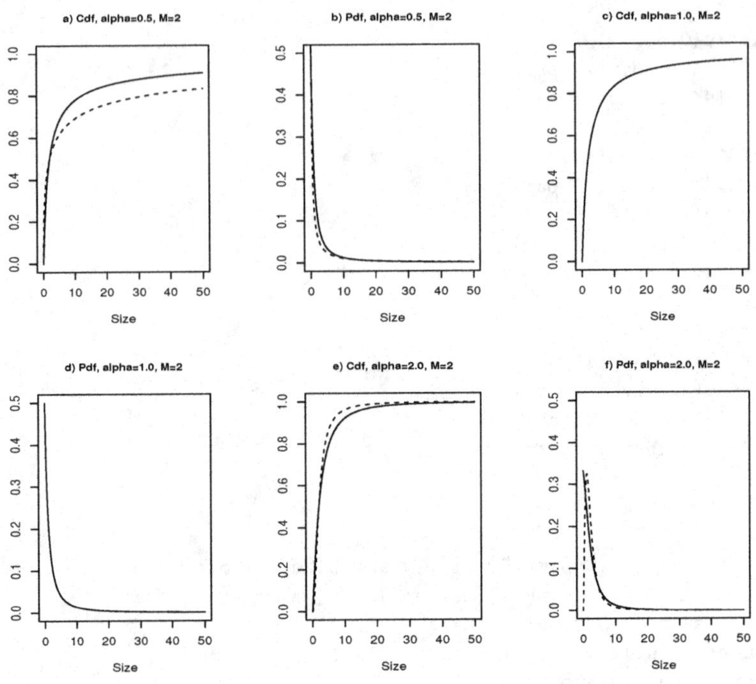

Figure 7.33 *Different shapes of the generalized Champernowne distribution. Positive c in solid line, and c* $= 0$ *in dashed line.*

```
xx=seq(0,5,length=2000)
u=seq(0,1,length=2000)

par.internal<-champPar(internal.data$y)
par.external<-champPar(external.data$y)

x0.internal=rep(0,length(internal.data$y))
y0.external=rep(0,length(external.data$y))

Den.parametric.internal<-dchamp(x,par.internal)
Den.parametric.external<-dchamp(x,par.external)

par(mfrow=c(2,2))
plot(x,Den.parametric.internal,type="l",col=1,xlab="Size",
     xlim=c(0,5),ylim=c(0,0.9),ylab="Probability",
     main='Internal GCD',cex.main=0.9,col.main=1)
points(internal.data$y,x0.internal,col=1)

plot(x,Den.parametric.internal,type="l",col=1,xlab="Size",
```

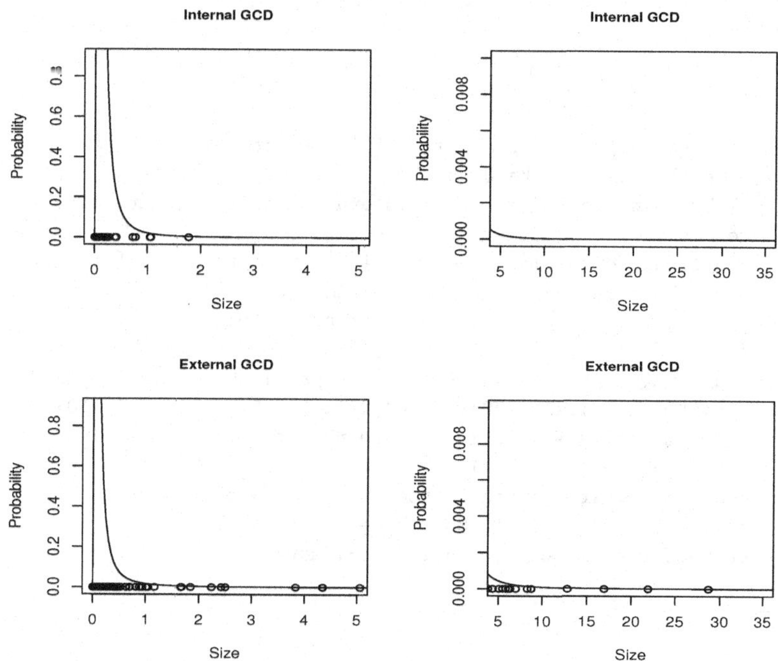

Figure 7.34 *The estimated density for internal example losses (above) and external example losses (below) operational risk data using the generalized Champernowne distribution (GCD).*

```
     xlim=c(5,35),ylim=c(0,0.01),ylab="Probability",
     main='Internal GCD',cex.main=0.9,col.main=1)
points(internal.data$y,x0.internal,col=1)

plot(x,Den.parametric.external,type="l",col=1,xlab="Size",
     xlim=c(0,5),ylim=c(0,0.9),ylab="Probability",
     main='External GCD',cex.main=0.9,col.main=1)
points(external.data$y,y0.external,col=1)

plot(x,Den.parametric.external,type="l",col=1,xlab="Size",
     xlim=c(5,35),ylim=c(0,0.01),ylab="Probability",
     main='External GCD',cex.main=0.9,col.main=1)
points(external.data$y,y0.external,col=1)


### Plotting Internal and External Quantile values for the
    distributions (Figure 2.7 and Figure 2.8)
```

```
q<-c(0.5,0.6,0.7,0.8,0.9,0.95,0.96,0.97,0.98,0.99)

# Internal

par(mfrow=c(2,2))
plot(q,qlogis(q,par.internal.L[[1]][1],par.internal.L[[1]][2]),
     type="b",col=1,xlab="Quantile",xlim=c(0.5,1),ylim=c(0,1),
     ylab="Quantile Value",main='Internal Logistic Quantiles',
     cex.main=0.9,col.main=1)
plot(q,qweibull(q,par.internal.W[[1]][1],par.internal.W[[1]][2]),
     type="b",col=1,xlab="Quantile",xlim=c(0.5,1),ylim=c(0,1),
     ylab="Quantile Value",main='Internal Weibull Quantiles',
     cex.main=0.9,col.main=1)
plot(q,qlnorm(q,par.internal.LN[[1]][1],par.internal.LN[[1]][2]),
     type="b",col=1,xlab="Quantile",xlim=c(0.5,1),ylim=c(0,10),
     ylab="Quantile Value",main='Internal Lognormal Quantiles',
     cex.main=0.9,col.main=1)
plot(q,qchamp(q,par.internal),type="b",col=1,xlab="Quantile",
     xlim=c(0.5,1),ylim=c(0,50),ylab="Quantile Value",
     main='Internal GCD Quantiles',cex.main=0.9,col.main=1)

# External

par(mfrow=c(2,2))
plot(q,qlogis(q,par.external.L[[1]][1],par.external.L[[1]][2]),
     type="b",col=1,xlab="Quantile",xlim=c(0.5,1),ylim=c(0,2),
     ylab="Quantile Value",main='External Logistic Quantiles',
     cex.main=0.9,col.main=1)
plot(q,qweibull(q,par.external.W[[1]][1],par.external.W[[1]][2]),
     type="b",col=1,xlab="Quantile",xlim=c(0.5,1),ylim=c(0,4),
     ylab="Quantile Value",main='External Weibull Quantiles',
     cex.main=0.9,col.main=1)
plot(q,qlnorm(q,par.external.LN[[1]][1],par.external.LN[[1]][2]),
     type="b",col=1,xlab="Quantile",xlim=c(0.5,1),ylim=c(0,10),
     ylab="Quantile Value",main='External Lognormal Quantiles',
     cex.main=0.9,col.main=1)
plot(q,qchamp(q,par.external),type="b",col=1,xlab="Quantile",
     xlim=c(0.5,1),ylim=c(0,205),ylab="Quantile Value",
     main='External GCD Quantiles',cex.main=0.9,col.main=1)
```

Figures in Chapter 3:

```
### Plotting the Classical Kernel Density Estimator (Figure 3.1)
# FUNCTIONS FOR KERNEL SMOOTHING
band = function(data){
```

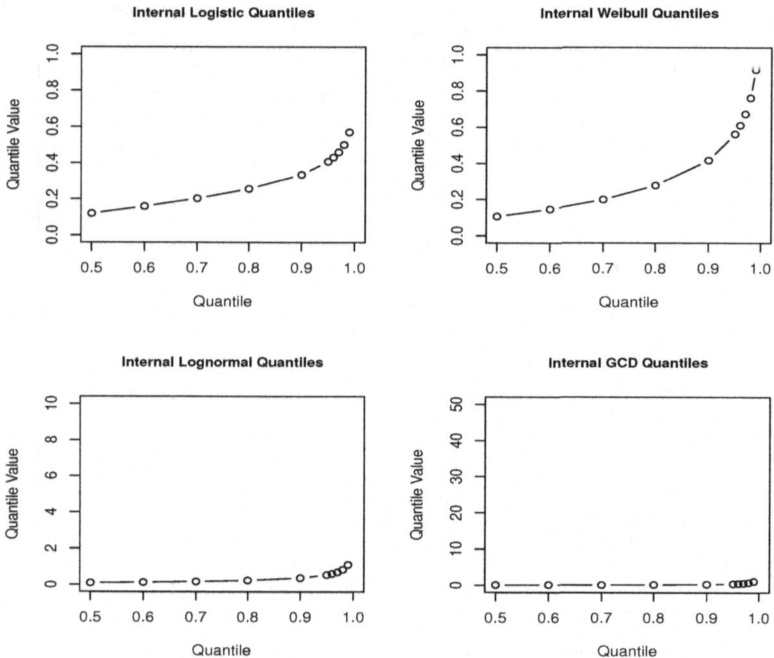

Figure 7.35 *Estimated quantile values from some distributions suggested in the example internal sample.*

```
((40*pi^(1/2))/length(data))^(1/5)*sd(data)
}


Epan = function(u){
0.75*(1-u^2)*(abs(u)<=1)
}
Grkorr01 = function(x,b){
# vre og nedre grnse i integralet for grnsekorrektionen

upperlim=pmin(1,x/b);
lowerlim=pmax(-1,(x-1)/b);
# Vrdien af den integrerede vrdi i vre og nedre grnse
KintUpper = 0.25*(3-upperlim^2)*upperlim;
KintLower = 0.25*(3-lowerlim^2)*lowerlim;
# Vrdien af grnsekorrektionen
kor = KintUpper-KintLower;
return(kor)
}
```

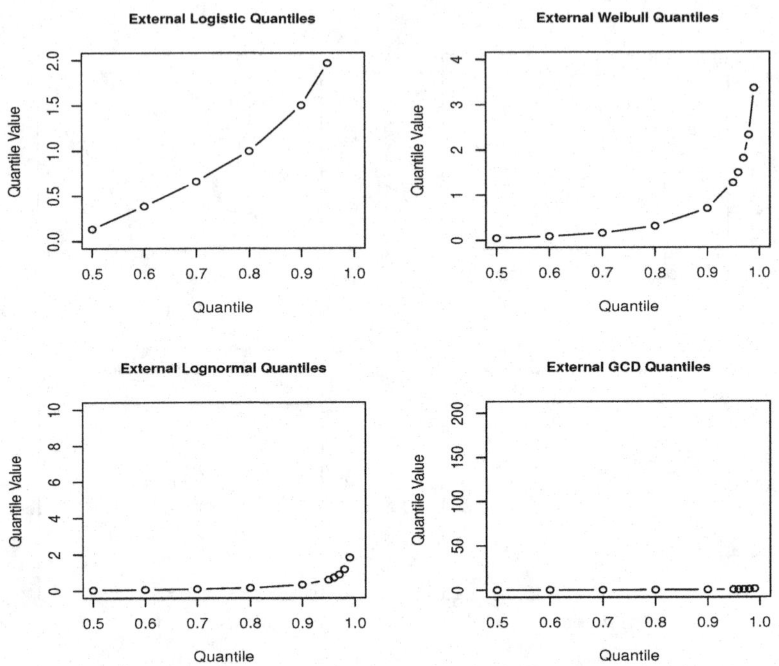

Figure 7.36 *Estimated quantile values from some distributions suggested in the example external sample.*

```
Grkorr02 = function(x,b){
# vre og nedre grnse i integralet for grnsekorrektionen

upperlim=pmin(1,x/b);
lowerlim=pmax(-1,(x-1)/b);
# Vrdien af den integrerede vrdi i vre og nedre grnse
KintUpper = 0.75*( 0.5-0.25*((upperlim)^2) )*(upperlim)^2;
KintLower = 0.75*( 0.5-0.25*((lowerlim)^2) )*(lowerlim)^2;
# Vrdien af grnsekorrektionen
kor = KintUpper-KintLower;
return(kor)
}


Grkorr03 = function(x,b){
# vre og nedre grnse i integralet for grnsekorrektionen

upperlim=pmin(1,x/b);
lowerlim=pmax(-1,(x-1)/b);
```

```
# Vrdien af den integrerede vrdi i vre og nedre grnse
KintUpper = 0.25*( 1-0.6*((upperlim)^2) )*((upperlim)^3);
KintLower = 0.25*( 1-0.6*((lowerlim)^2) )*((lowerlim)^3);
# Vrdien af grnsekorrektionen
kor = KintUpper-KintLower;
return(kor)
}


Grkorr04 = function(x,b){
# vre og nedre grnse i integralet for grnsekorrektionen

upperlim=pmin(1,x/b);
lowerlim=pmax(-1,(x-1)/b);
# Vrdien af den integrerede vrdi i vre og nedre grnse
KintUpper = 0.75*( 0.25-(1/6)*((upperlim)^2) )*((upperlim)^4);
KintLower = 0.75*( 0.25-(1/6)*((lowerlim)^2) )*((lowerlim)^4);
# Vrdien af grnsekorrektionen
kor = KintUpper-KintLower;
return(kor)
}
KerDenClassical = function(x, data, band){
#       Inputs:     data    : Transformed Data
#                   x       : vector between [0,1]
#   band    : bandwidth
#
#       Outputs:    Den     : Density estimator
#                   Quan    : Quantile estimator
#                   Control : Integration value

Grkorr =Grkorr01

# Endimensional tthed
b = band
xM = rep(1,times=length(data)) %o% x
dataM = data %o% rep(1,times=length(x))
arg = Epan((xM-dataM)/b)
Density = colSums(arg)
Density = Density / (length(data)*b)

# Change made for OR-II article: changed splinefun to approxfun
Densityapprox = splinefun(x, Density)

Control = integrate(Densityapprox,0,1)[[1]]

# Density function
Den = function(x)
```

```
{
Densityapprox(x)
}

# Quantile function
Integrant = function(m,q)
{
integrate(Den,0,m)[[1]]-q
}

Quan = function(q)
{
uniroot(function (m) Integrant(m,q), c(0,1))$root
}

  return(Den)
}


band.int<-band(internal.data) #0.1241
band.int.1<-0.1
band.int.2<-0.5
band.int.3<-1

band.ext<-band(external.data) #1.1735
band.ext.1<-1
band.ext.2<-2
band.ext.3<-3

KerDenClassical.internal.1<-KerDenClassical(x,internal.data$y,
                                            band.int.1)
KerDenClassical.external.1<-KerDenClassical(x,external.data$y,
                                            band.ext.1)

KerDenClassical.internal.2<-KerDenClassical(x,internal.data$y,
                                            band.int.2)
KerDenClassical.external.2<-KerDenClassical(x,external.data$y,
                                            band.ext.2)

KerDenClassical.internal.3<-KerDenClassical(x,internal.data$
                                            y,band.int.3)
KerDenClassical.external.3<-KerDenClassical(x,external.data$
                                            y,band.ext.3)

par(mfrow=c(2,2))
plot(x,KerDenClassical.internal.1(x),type="l",col=1,xlab="Size",
     xlim=c(0,5),ylim=c(0,0.9),ylab="Probability",
```

```
     main='Internal Classical KDE',cex.main=0.9,col.main=1)
points(internal.data$y,x0.internal,col=1)
lines(x,KerDenClassical.internal.2(x),lty=2)
lines(x,KerDenClassical.internal.3(x),lty=3)

plot(x,KerDenClassical.internal.1(x),type="l",col=1,xlab="Size",
     xlim=c(5,35),ylim=c(0,0.01),ylab="Probability",
     main='Internal Classical KDE',cex.main=0.9,col.main=1)
points(internal.data$y,x0.internal,col=1)
lines(x,KerDenClassical.internal.2(x),lty=2)
lines(x,KerDenClassical.internal.3(x),lty=3)
,cex=0.8,lty=c(1,2,3),bty="n",col=c(1,1,1))

plot(x,KerDenClassical.external.1(x),type="l",col=1,xlab="Size",
     xlim=c(0,5),ylim=c(0,0.9),ylab="Probability",
     main='External Classical KDE',cex.main=0.9,col.main=1)
points(external.data$y,y0.external,col=1)
lines(x,KerDenClassical.external.2(x),lty=2)
lines(x,KerDenClassical.external.3(x),lty=3)

plot(x,KerDenClassical.external.1(x),type="l",col=1,xlab="Size",
     xlim=c(5,35),ylim=c(0,0.01),ylab="Probability",
     main='External Classical KDE',cex.main=0.9,col.main=1)
points(external.data$y,y0.external,col=1)
lines(x,KerDenClassical.external.2(x),lty=2)
lines(x,KerDenClassical.external.3(x),lty=3)$


### Plotting the Transformation (Champernowne) Local Constant
    Kernel Density Estimator Process (Figures 3.2 and 3.3) [0,1]

KerDen = function(x, data, band){
#       Inputs:     data    : Transformered Data
#                   x       : vector between [0,1]
#   band     : bandwidth
#
#       Outputs:    Den     : Density estimator
#                   Quan    : Quantile estimator
#                   Control : Integration value

Grkorr =Grkorr01

# Endimensional tthed
b = band
xM = rep(1,times=length(data)) %o% x
dataM = data %o% rep(1,times=length(x))
arg = Epan((xM-dataM)/b)
Density = colSums(arg)
```

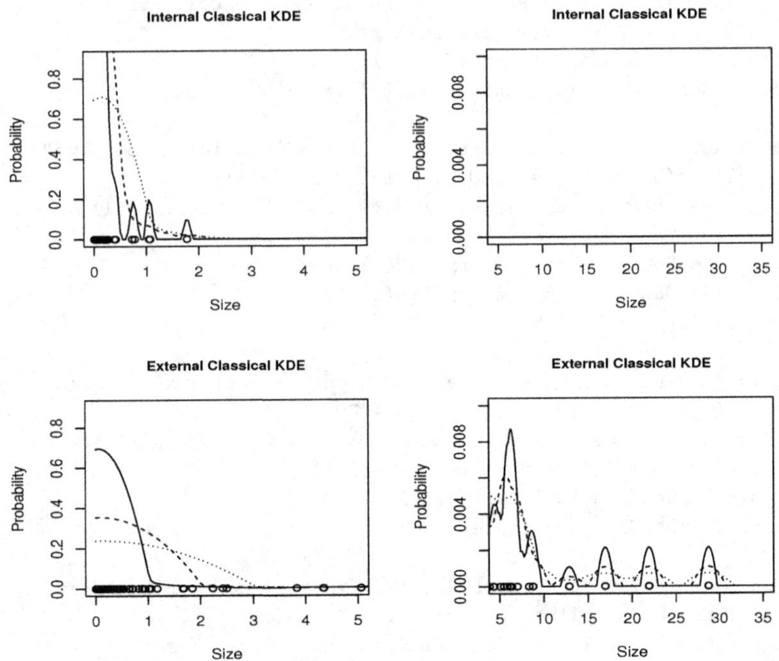

Figure 7.37 *Classical kernel density estimation for the example operational risk data sets. Above (below) the internal (external) operational risk data are plotted. Right-hand-side plots focus on large losses with adequate vertical scale units. Solid line corresponds to a bandwidth of size* $b = 0.1$, *dashed line corresponds to* $b = 0.5$, *and dotted line corresponds to* $b = 1$.

```
Density = Density / (Grkorr(x,b)*length(data)*b)

# Change made for OR-II article: changed splinefun to approxfun
Densityapprox = splinefun(x, Density)

Control = integrate(Densityapprox,0,1)[[1]]

# Density function
Den = function(x)
{
Densityapprox(x)/Control
}

# Quantile function
Integrant = function(m,q)
{
```

```
integrate(Den,0,m)[[1]]-q
}

Quan = function(q)
{
uniroot(function (m) Integrant(m,q), c(0,1))$root
}

  return(Den)
}
pchamp = function(x,par)
{
a=par[1]
M=par[2]
c=par[3]
((x+c)^a-c^a)/((x+c)^a+(M+c)^a-2*c^a)
}
dchamp = function(x,par)
{
a=par[1]
M=par[2]
c=par[3]
(a*(x+c)^(a-1)*((M+c)^a-c^a))/((x+c)^a+(M+c)^a-2*c^a)^2
}


qchamp = function(y,par)
{
a=par[1]
M=par[2]
c=par[3]
((y*(M+c)^a+(1-2*y)*c^a)/(1-y))^(1/a)-c
}
champLike = function(data,par)
{
a=par[1]
M=par[2]
c=par[3]
N = length(data)
N*log(a)+N*log((M+c)^a-c^a)+(a-1)*sum(log(data+c))
    -2*sum(log((data+c)^a+(M+c)^a-2*c^a))
}

champPar = function(data)
{
N = length(data)
par = matrix(0,nrow=3,ncol=1)
```

```
L = 0
avalg = 0
Lvalg = 0
cvalg = seq(0,2,0.5)
par[2] = median(data)
seqp = seq(1,3,1)
seqj = seq(1,21,1)
# Finder optimalt a for hver vrdi af c
for (i in seq(1,length(cvalg),1))
{
par[3]=cvalg[i]*par[2]
astart=0.1
aslut=21
for (praecision in seqp)
{
for (j in seqj)
{
par[1]=astart+(j-1)*(aslut-astart)/20
L[j]=champLike(data,par)
}
jmax=which.max(L)
astart=max(0.0001,astart+(jmax-2)*(aslut-astart)/20)
aslut=astart+jmax*(aslut-astart)/20
}
avalg[i]=astart+(jmax-1)*(aslut-astart)/20
Lvalg[i]=L[jmax]
}
iopt=which.max(Lvalg)
par[1]=avalg[iopt]
par[3]=cvalg[iopt]*par[2]
return(par)
}
par.internal<-champPar(internal.data$y)
par.external<-champPar(external.data$y)

# Transform Data
internal.data.tr<-pchamp(internal.data$y,par.internal)
external.data.tr<-pchamp(external.data$y,par.external)

# KDE
KerDen.internal<-KerDen(u,internal.data.tr,
                        band(internal.data.tr))
KerDen.external<-KerDen(u,external.data.tr,
                        band(external.data.tr))


par(mfrow=c(1,2))
```

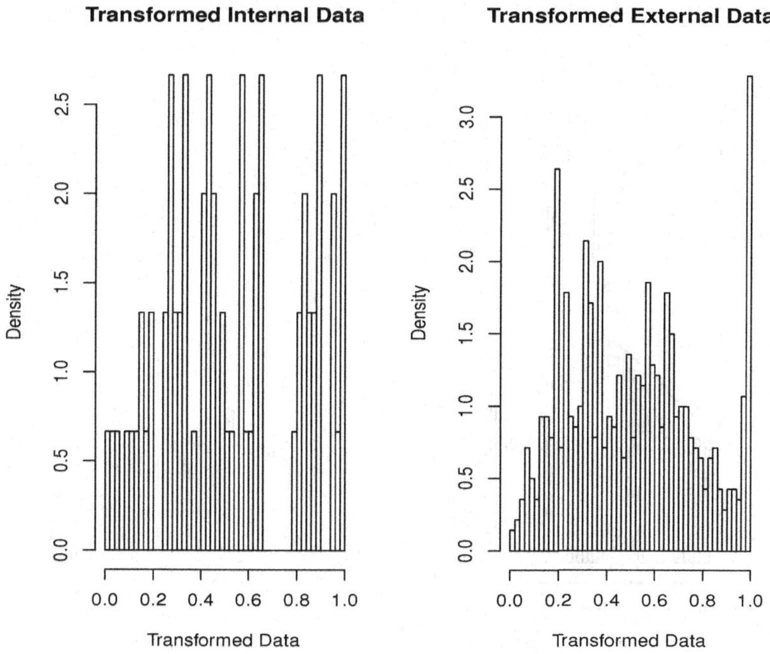

Figure 7.38 *Transformed example internal and example external operational risk data with the estimated cdf of the generalized Champernowne distribution.*

```
hist(internal.data.tr,nclass=60,freq=0,xlab="Transformed Data",
     main="Transformed Internal Data")
hist(external.data.tr,nclass=60,freq=0,xlab="Transformed Data",
     main="Transformed External Data")


par(mfrow=c(1,2))
hist(internal.data.tr,nclass=60,freq=0,xlab="Transformed Data",
     main="KDE on Transformed Internal Data ")
lines(u,KerDen.internal(u))
hist(external.data.tr,nclass=60,freq=0,xlab="Transformed Data",
     main="KDE on Transformed External Data")
lines(u,KerDen.external(u))
$



### Plotting the Transformation (Champernowne) Local Constant
```

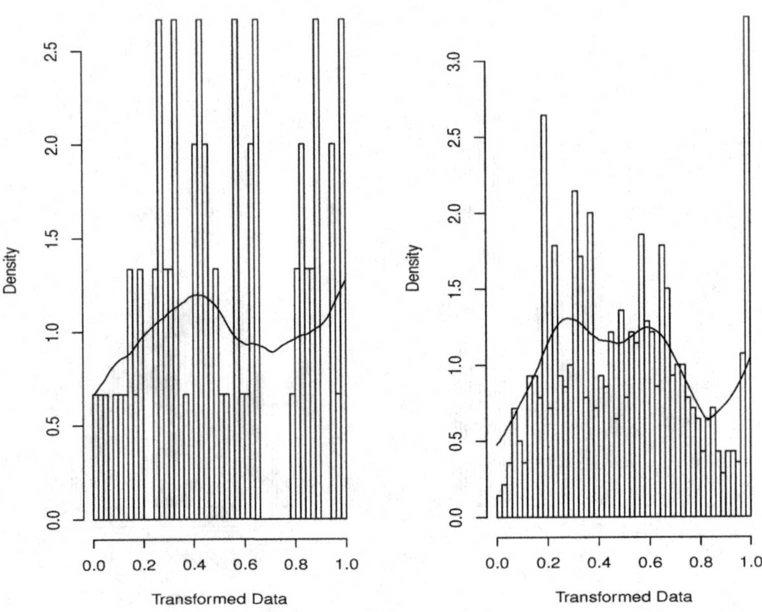

Figure 7.39 *Transformed example internal (left) and example external (right) operational risk data with the generalized Champernowne distribution and the respective kernel density estimation (KDE) with local constant boundary correction.*

```
    Kernel Density Estimator (Figure 3.4) [0,Inf)

KerDenLL = function(x, data, band){
#       Inputs:     data    : Transformed Data
#                   x       : vector between [0,1]
#
#       Outputs:    Den     : Density estimator
#                   Quan    : Quantile estimator
#                   Control : Integration value

Grkorr =Grkorr01
Grkorr2 =Grkorr02
Grkorr3 =Grkorr03


# Endimensional tthed
b = band;
xM = rep(1,times=length(data)) %o% x
```

```
dataM = data %o% rep(1,times=length(x))

arg = Epan((xM-dataM)/b)
argL = (((xM-dataM)/b))*Epan((xM-dataM)/b)

Density = colSums(arg)
DensityL = colSums(argL)

Density = (Density*Grkorr3(x,b) )  /
              ((Grkorr3(x,b)*Grkorr(x,b)-(Grkorr2(x,b))^2)*
                 length(data)*b )
DensityL =(DensityL*(-Grkorr2(x,b)) ) /
              ((Grkorr3(x,b)*Grkorr(x,b)-(Grkorr2(x,b))^2)*
                length(data)*b )

DensityLL= Density+DensityL

# Change made for OR-II article: changed splinefun to approxfun
Densityapprox = approxfun(x, DensityLL)

Control = integrate(Densityapprox,0,1)[[1]]

# Density function
Den = function(x)
{
Densityapprox(x)/Control
}

# Quantile function
Integrant = function(m,q)
{
integrate(Den,0,m)[[1]]-q
}

Quan = function(q)
{
uniroot(function (m) Integrant(m,q), c(0,1))$root
}

  return(Den, Quan, Control)
}


TrKerDen = function(x, data, distr, par, band, kernel){
# Input: x : vector between [0,Inf)
# data    : Original Data
# distr : Distribution "champ", "lnorm" or "weibull"
```

```
# par     : Estimated Parameters of the 'distr'
# band    : bandwidth (on transformed data)
# kernel : Local Constant ("LC") or Local Linear ("LL") kernel
#
# Output: Den      : Density estimator
# # Quan     : Quantile estimator
# Control : Integration value

if(distr == "champ"){
Trans = function(x,par) pchamp(x,par)
dataTrans = pchamp(data,par)
# Transinv =  qchamp
trans = function(x,par) dchamp(x,par)
}
if(distr == "lnorm"){
Trans = function(x,par) plnorm(x, meanlog = par[1],
        sdlog = par[2], lower.tail = TRUE, log.p = FALSE)
dataTrans = plnorm(data,meanlog = par[1],
        sdlog = par[2], lower.tail = TRUE, log.p = FALSE)
# Transinv =  qlnorm(p, meanlog = par[1],
        sdlog = par[2], lower.tail = TRUE, log.p = FALSE)
trans = function(x,par) dlnorm(x,meanlog = par[1],
        sdlog = par[2], log = FALSE)
}
if(distr == "weibull"){
Trans = function(x,par) pweibull(x,shape = par[1],
                                scale = par[2],
                                lower.tail = TRUE,
                                log.p = FALSE)
dataTrans = pweibull(data,  shape = par[1], scale = par[2],
                            lower.tail = TRUE, log.p = FALSE)
# Transinv =  qweibull(p,shape = par[1], scale = par[2],
                            lower.tail = TRUE, log.p = FALSE)
trans = function(x,par) dweibull(x,shape = par[1],
                                scale = par[2], log = FALSE)
}

xTrans = c(Trans(x,par),1)

if(kernel == "LC") TransEst = KerDen(xTrans,dataTrans, band)
if(kernel == "LL") TransEst = KerDenLL(xTrans,dataTrans, band)

 #Control = TransEst$Control

# Transformer til original akse
Den = function(x)
{
```

```
TransEst(Trans(x,par)) * trans(x,par)
}

return(Den)
}
Den.parametric.internal<-dchamp(x,par.internal)
Den.parametric.external<-dchamp(x,par.external)

KerDen.internal.tr<-TrKerDen(x,internal.data$y,"champ",
                             par.internal,
                             band(internal.data.tr),"LC")
KerDen.external.tr<-TrKerDen(x,external.data$y,"champ",
                             par.external,
                             band(external.data.tr),"LC")

par(mfrow=c(2,2))
plot(x,KerDen.internal.tr(x),type="l",col=1,xlab="Size",
     xlim=c(0,5),ylim=c(0,0.9),ylab="Probability",
     main='Estimated KDE for Internal Data',cex.main=0.9,
     col.main=1)
points(internal.data$y,x0.internal,col=1)
lines(x,Den.parametric.internal,lty=2)

plot(x,KerDen.internal.tr(x),type="l",col=1,xlab="Size",
     xlim=c(5,35),ylim=c(0,0.01),ylab="Probability",
     main='Estimated KDE for Internal Data',cex.main=0.9,
     col.main=1)
points(internal.data$y,x0.internal,col=1)
lines(x,Den.parametric.internal,lty=2)

plot(x,KerDen.external.tr(x),type="l",col=1,xlab="Size",
     xlim=c(0,5),ylim=c(0,0.9),ylab="Probability",
     main='Estimated KDE for External Data',cex.main=0.9,
     col.main=1)
points(external.data$y,y0.external,col=1)
lines(x,Den.parametric.external,lty=2)

plot(x,KerDen.external.tr(x),type="l",col=1,xlab="Size",
     xlim=c(5,35),ylim=c(0,0.01),ylab="Probability",
     main='Estimated KDE for External Data',cex.main=0.9,
     col.main=1)
points(external.data$y,y0.external,col=1)
lines(x,Den.parametric.external,lty=2)
```

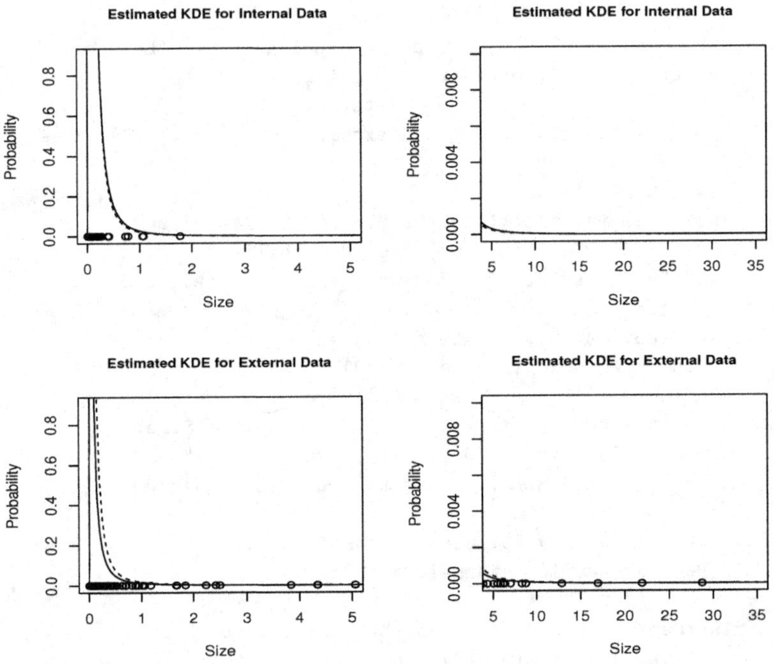

Figure 7.40 *Example internal (above) and example external (below) operational risk data in the original scale. Solid line represents the transformation semiparametric kernel estimation with the generalized Champernowne distribution and a local constant boundary correction. The dashed line corresponds to a pure parametric fit of a generalized Champernowne distribution. Right-hand-side plots focus on large losses with adequate vertical scale units.*

Bibliography

[1] C. Alexander. *Risk: Regulation, Analysis and Management.* Pearson Education, Upper Saddle River, NJ, 2003.

[2] A. Bangia, F. X. Diebold, T. Schuermann, and J. Stroughair. Modeling liquidity risk, with implications for traditional market risk measurement and management. In *Risk Management: The State of the Art.* Kluwer Academic Publishers, Dordrecht, Netherlands, 2001.

[3] N. Baud, A. Frachot, and T. Roncalli. Internal data, external data and consortium data for operational risk measurement: how to pool data properly. Working paper, Groupe de Recherche Opérationnelle, Crédit Lyonnais, 2002.

[4] C. Bolancé, M. Guillén, and J. P. Nielsen. Kernel density estimation of actuarial loss functions. *Insurance: Mathematics and Economics,* 32(1):19–36, 2003.

[5] C. Bolancé, M. Guillén, and J. P. Nielsen. Inverse Beta transformation in kernel density estimation. *Statistics & Probability Letters*, 78:1757–1764, 2008.

[6] C. Bolancé, M. Guillén, and J. P. Nielsen. Transformation kernel estimation of insurance claim cost distributions. In *Mathematical and Statistical Methods for Actuarial Sciences and Finance*, pages 43–51. Corazza, M. and Pizzi, C. (Eds), Springer, Berlin, 2010.

[7] T. Buch-Kromann, M. Englund, J. Gustafsson, J. P. Nielsen, and F. Thuring. Non-parametric estimation of operational risk losses adjusted for under-reporting. *Scandinavian Actuarial Journal*, 4:293–304, 2007.

[8] T. Buch-Kromann, M. Guillén, J. P. Nielsen, and O. Linton. Multivariate density estimation using dimension reducing information and tail flattening transformations. *Insurance: Mathematics and Economics*, 48(1):99–110, 2011.

[9] T. Buch-Larsen. A unified approach to the estimation of financial and actuarial loss distributions. Master thesis, Laboratory of Actuarial Mathematics, University of Copenhagen, 2003.

[10] T. Buch-Larsen, J. P. Nielsen, M. Guillén, and C. Bolancé. Kernel density estimation for heavy-tailed distributions using the Champernowne

transformation. *Statistics*, 39(6):503–518, 2005.

[11] H. Bühlmann, P. V. Shevchenko, and M. V. Wüthrich. A "Toy" Model for Operational Risk Quantification using Credibility Theory. *The Journal of Operational Risk*, 2:3–19, 2006.

[12] H. Bühlmann, P.V. Shevchenko, and M.V. Wüthrich. A "toy" model for operational risk quantification using credibility theory. *Journal of Operational Risk*, 2(1):3–20, 2007.

[13] Bolancé C., M. Guillén, E. Pelican, and R. Vernic. Skewed bivariate models and nonparametric estimation for the cte risk measure. *Insurance: Mathematics and Economics*, 43(3):386 – 393, 2008.

[14] D. G. Champernowne. The Oxford Meeting, September 25–29. *Econometrica*, 5(October 1937), 1936.

[15] D. G. Champernowne. The graduation of income distributions. *Econometrica*, 20:591–615, 1952.

[16] V. Chavez-Demoulin, P. Embrechts, and J. J. Nešlehová. Quantitative models for operational risk: Extremes, dependence and aggregation. *Journal of Banking and Finance*, 30(10):26352658, 2006.

[17] A. S. Chernobai, S. T. Rachev, and F.J. Fabozzi. *Operational Risk: A Guide to Basel II Capital Requirements, Modelling and Analysis.* John Wiley & Sons, Hoboken, NJ, 2007.

[18] P. Cizek, W. Härdle, and R. Weron. *Statistical Tools for Finance and Insurance.* Springer, Berlin, 2005.

[19] R.T. Clemen and R L. Winkler. Combining probability distributions from experts in risk analysis. *Risk Analysis*, 19(2):187–203, 1999.

[20] A. E. Clements, A. S. Hurn, and K. A. Lindsay. Möbius-like mappings and their use in kernel density estimation. *Journal of the American Statistical Association*, 98(464):993–1000, 2003.

[21] S. G. Coles. *An Introduction to Statistical Modelling of Extreme Values.* Springer, Berlin, 2001.

[22] M. G. Cruz. *Modeling, Measuring and Hedging Operational Risk.* John Wiley & Sons, Hoboken, NJ, 2002.

[23] H. Dahen and G. Dionne. Scaling models for the severity and frequency of external operational loss data. *Journal of Banking & Finance*, 34(7):1484–1496, 2007.

[24] E. Davis. *The Advanced Measurement Approach to Operational Risk.* Risk Books, London, 2006.

[25] C. de Boor. *A Practical Guide to Splines.* Volume 27, Springer, Berlin, 2001.

[26] P. de Fountnouvelle, V. De Jesus-Rueff, J. Jordan, and E. Rosengren. Using loss data to quantify operational risk. Technical report, Technical

report, Federal Reserve Bank of Boston and Fitch Risk, 2003.

[27] M. Degen, P. Embrechts, and D. D. Lambrigger. The quantitative modeling of operational risk. Between g-and-h and EVT. *Astin Bulletin*, 37(2):265–292, 2007.

[28] K. Dutta and J. Perry. A tale of tails: an empirical analysis of loss distribution models for estimating operational risk capital. http://www.bos.frb.org/economic/wp/wp2006/wp0613.htm, 2006.

[29] P. Embrechts. *Extremes and Integrated Risk Management.* Risk Books, London, 2000.

[30] P. Embrechts and M. Hofert. Practices and issues in operational risk modeling under Basel II. *Lithuanian Mathematical Journal*, 51(2):180193, 2011.

[31] P. Embrechts, C. Kluppelberg, and T. Mikosch. *Modelling extremal events for insurance and finance.* Springer, Berlin, 1997.

[32] P. Embrechts and G. Puccetti. Aggregating risk capital, with an application to operational risk. *Geneva Risk and Insurance Review*, 30(2):71–90, 2006.

[33] P. Embrechts and G. Puccetti. Aggregating operational risk across matrix structured loss data. *Jurnal of Operational Risk*, 3(2):29–44, 2008.

[34] J. Fan and I. Gijbels. *Local Polynomial Modelling and Its Applications.* Chapman & Hall, London, 1996.

[35] S. Figini, P. Giudici, P. Uberti, and A. Sanyal. A statistical method to optimize the combination of internal and external data in operational risk measurement. *Journal of Operational Risk*, 2(4):69–78, 2008.

[36] A. Frachot and T. Roncalli. Mixing internal and external data for managing operational risk. Working paper, Groupe de Recherche Opérationnelle, Crédit Lyonnais, 2002.

[37] C. Franzetti. *Operational Risk Modelling and Management.* CRC Press, New York, 2010.

[38] E. Frees and E. Valdez. Understanding relationships using copulas. *North American Actuarial Journal*, 2(1):1–25, 1998.

[39] J. Gavin, S. Haberman, and R. Verrall. Moving weighted average graduation using kernel estimation. *Insurance, Mathematics and Economics*, 12(2):113–126, 1993.

[40] J. Gavin, S. Haberman, and R. Verrall. On the choice of bandwidth for kernel graduation. *Journal of the Institute of Actuaries*, 121(478):119–134, 1994.

[41] R. Giacometti, S. Rachev, A. Chernobai, and M. Bertochi. Aggregation issues in operational risk. *Journal of Operational Risk*, 3(3), 2008.

[42] M. Guillén, J. Gustafsson, and J. P. Nielsen. Combining underreported

internal and external data for operational risk measurement. *Journal of Operational Risk*, 3(4):3–24, 2008.

[43] M. Guillén, J. Gustafsson, J. P. Nielsen, and P. Pritchard. Using external data in operational risk. *Geneva Papers on Risk and Insurance—Issues and Practice*, 32(2):178–189, 2007.

[44] M. Guillén, F. Prieto, and J. M. Sarabia. Modelling losses and locating the tail with the pareto positive stable distribution. *Insurance: Mathematics and Economics*, 49(3):454–461, 2011.

[45] J. Gustafsson. A mixing severity model incorporating three sources of data for operational risk quantification. *Insurance Markets and Companies: Analyses and Actuarial Computations*, 2010/1:54–68, 2010.

[46] J. Gustafsson, M. Hagmann, J. P. Nielsen, and O. Scaillet. Local transformation kernel density estimation of loss distributions. *Journal of Business and Economic Statistics*, 27:161–175, 2009.

[47] J. Gustafsson and J. P. Nielsen. A mixing model for operational risk. *The Journal of Operational Risk*, 3(3):25–37, 2008.

[48] J. Gustafsson, J. P. Nielsen, P. Pritchard, and D. Roberts. Quantifying operational risk guided by kernel smoothing and continuous credibility. *The Journal of Operational Risk*, 2006.

[49] J. Gustafsson, F. Thuring, and P. Pritchard. A suitable parametric model for operational risk applications. Unpublished.

[50] N. L. Hjort and I. K. Glad. Nonparametric density estimation with a parametric start. *The Annals of Statistics*, 23(2):882–904, 1995.

[51] O. Hössjer and D. Ruppert. Asymptotics for the transformation kernel density estimator. *The Annals of Statistics*, 23(4):1198–1222, 1995.

[52] N. Johnson, S. Kotz, and N. Balakrishnan. *Continuous Univariate Distributions, Vol 1*. John Wiley & Sons, New York, 2nd edition, 1994.

[53] N. Johnson, S. Kotz, and N. Balakrishnan. *Continuous Univariate Distributions, Vol 2*. John Wiley & Sons, New York, 2nd edition, 1995.

[54] M.C. Jones, O. Linton, and J. P. Nielsen. A simple bias reduction method for density estimation. *Biometrika*, 82(2):327–338, 1995.

[55] J. L. King. *Operational Risk: Measurement and Modelling*. John Wiley & Sons, New York, 2001.

[56] S. A. Klugman, H. A. Panjer, and G. E. Willmot. *Loss Models: From Data to Decisions*. John Wiley & Sons, New York, 1998.

[57] S. A. Klugman, H. A. Panjer, and G. E. Willmot. *Loss Models: From Data to Decisions*. John Wiley & Sons, New York, 2008.

[58] S. Kotz and S. Nadarajah. *Extreme Value Distributions: Theory and Applications*. Imperial College Press, London, 2000.

[59] D. D. Lambrigger, P. V. Shevchenko, and M. V. Wüthrich. The quantification of operational risk using internal data, relevant external data and expert opinion. *Journal of Operational Risk*, 2(3):3–27, 2007.

[60] A.J. McNeil, R. Frey, and P. Embrechts. *Quantitative Risk Management: Concepts, Techniques and Tools*. Princeton Series in Finance, Princenton University Press, Princenton, NJ, 2005.

[61] J. Nešlehová, P. Embrechts, and V. Chavez-Demoulin. Infinite mean models and the LDA for operational risk. *Journal of Operational Risk*, 1(1):3–25, 2006.

[62] H. H. Panjer. *Operational Risk: Modeling Analytics*. John Wiley & Sons, New York, 2006.

[63] M. Power. The invention of operational risk. *Review of International Political Economy*, 4(12):577–599, 2005.

[64] R. D. Reiss and M. Thomas. *Statistical Analysis of Extreme Values with Applications to Insurance, Finance, Hydrology and Other Fields*. Birkhauser-Verlag, Cambridge, MA, 2001.

[65] S. I. Resnick. *Extreme Values, Regular Variation, and Point Processes*. Springer, Berlin, 2007.

[66] D. Ruppert and D. B. H. Cline. Bias reduction in kernel density estimation by smoothed empirical transformations. *The Annals of Statistics*, 22(1):185–210, 1994.

[67] J. M. Sarabia and M. Guillén. Joint modelling of the total amount and the number of claims by conditionals. *Insurance: Mathematics and Economics*, 43(3):466 – 473, 2008.

[68] J. M. Sarabia and F. Prieto. The pareto-positive stable distribution: A new descriptive model for city size data. *Physica A: Statistical Mechanics and its Applications*, 388(19):4179 – 4191, 2009.

[69] P. V. Shevchenko and M. V. Wüthrich. The structural modeling of operational risk via Bayesian inference. *Journal of Operational Risk*, 1(3):3–26, 2006.

[70] B. W. Silverman. *Density Estimation for Statistics and Data Analysis*. Chapman and Hall, London, 1986.

[71] M. H. Tripp, H. L. Bradley, R. Devitt, G. C. Orros, G. L. Overton, L. M. Pryor, and R. A. Shaw. Quantifying operational risk in general insurance companies. *British Actuarial Journal*, 10:919–1012, 2004.

[72] R.J. Verrall, R. Cowell, and Y. Y. Khoon. Modelling Operational Risk with Bayesian Networks. *Journal of Risk and Insurance*, 74:795–827, 2007.

[73] M. P. Wand and M. C. Jones. *Kernel Smoothing*. Chapman and Hall, London, 1995.

[74] M. P. Wand, J. S. Marron, and D. Ruppert. Transformation in density estimation (with comments). *Journal of the American Statistical Association*, 86(414):343–361, 1991.

[75] R. Wei. Quantification of operational losses using firm-specific information and external databases. *Journal of Operational Risk*, 1(4):3–34, 2007.

[76] L. Yang. Root-n convergent transformation-kernel density estimation. *Journal of Nonparametric Statistics*, 12(4):447–474, 2000.

[77] L. Yang and J. S. Marron. Iterated transformation-kernel density estimation. *Journal of the American Statistical Association*, 94(446):580–589, 1999.

Index

Printed in the United States
by Baker & Taylor Publisher Services